THE BIG DIG

THE BIG DIG

by Dan McNichol

Photographs by Andy Ryan

To my Dad, the greatest builder I have ever known.
His passion to create has always inspired me.
—Dan McNichol

Editorial Director:	Barbara J. Morgan
Editor:	Marjorie Palmer
Design:	Richard J. Berenson
	Berenson Design & Books, Ltd., New York, NY
Production:	Della R. Mancuso
	Mancuso Associates, Inc., North Salem, NY

Library of Congress Cataloging-in-Publication Data is available on request.

ISBN 0-7607-2307-9

Printed in Italy

9 8 7 6 5 4 3 2

Second Edition

live in Florida now and don't get up to Boston very much these days, but I still follow news about the city and what's going on up there. I know they're still working on the Big Dig, but don't worry, they'll finish it. They talked for fifty years about the tunnel they named after me, but they finally got it done and I hear it's working out. Boston's the greatest city on earth, and the people and fans there will always be first in my heart.

——Ted Williams

CHAPTER ONE
What Is The Big Dig?
Page 8

CHAPTER TWO
Boston's Dirty History
Page 16

CHAPTER THREE
All Politics Are Local
Page 26

CHAPTER FOUR
Gearing Up
Page 40

CHAPTER FIVE
The Ted Williams Tunnel
Page 52

CHAPTER SIX
Spectacle Island
Page 84

CHAPTER SEVEN
Slurry Walls and the Maze Below
Page 104

CHAPTER EIGHT
Underground in Downtown
Page 118

CHAPTER NINE
The Mother of All Interchanges
Page 154

CHAPTER TEN
The Big Dig's Biggest Challenge
Page 172

CHAPTER ELEVEN
The Charles River Crossings
Page 198

CHAPTER TWELVE
The Home Stretch
Page 214

Acknowledgments & Credits 232 • Big Dig Timeline 234 • Unions 236 • Index 237

CHAPTER ONE
What Is The Big Dig?

"IT'S JUST A LITTLE ROAD PROJECT up here in Boston." That's the way Peter M. Zuk, then project director for the Central Artery/Tunnel Project, described the Big Dig on National Public Radio in 1997.

In fact, the Central Artery/Tunnel Project, known more colorfully as the Big Dig, is the largest and most complex urban infrastructure project ever undertaken in the modern world. It is bigger in scale than the Panama Canal or Hoover Dam and more complex in its planning, engineering, and construction than the two combined. Its innovative solutions to a host of complicated problems have advanced engineering around the world.

In a nutshell, the Big Dig is the reconstruction and construction of two sections of interstate highway, nearly eight linear miles, running through the heart of Boston. Half of these miles are underground or underwater, tunneling beneath the mucky bottoms of harbors, dodging a maze of subway lines, building foundations, old wharves, and other obstacles. Many parts of the highway are 12 lanes wide. That works out to 161 lane miles of superhighway in tunnels, bridges, viaducts, and surface roads.

The route of the project travels directly through Boston's downtown, where nearly five miles of I-93 are being reconstructed from the Big Dig's southernmost point in Roxbury to the northern tip in Charlestown. At the northern end of the project, an old, severely strained trestle bridge is being replaced by the world's widest cable-stayed bridge, spanning the Charles River and creating a new focal point in the city skyline.

Interstate 90's extension to Logan International Airport is the other part of the Big Dig's undertaking. I-90 is the longest interstate in the United States, extending from Seattle, Washington, to Boston, Massachusetts. It was intended to be coast-to-coast across America, part of a defense plan of networking highways conceived by President Dwight D. Eisenhower in the 1950s. However, it now dead-ends into I-93, three miles short of its intended destination, the airport.

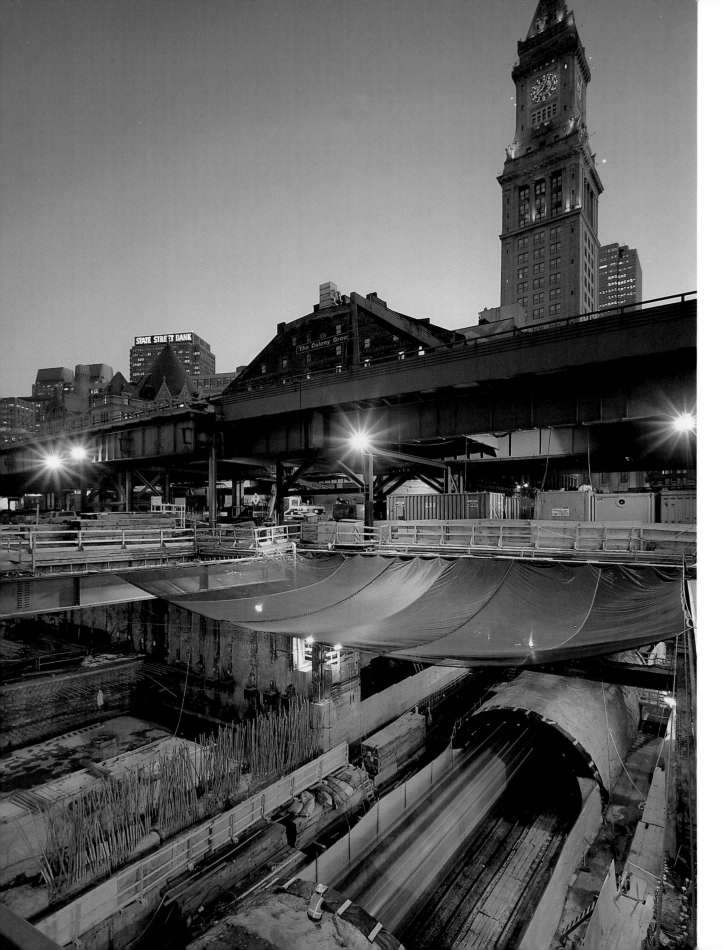

Spring of 2000, the author is perched upon the country's first underwater tunnel to carry a train, Boston's Blue Line Subway. Ten Lanes of highway tunnel are being built over the top of the subway. The elevated highway will be torn down in 2004. Boston's Custom House Tower looms in the background.

11

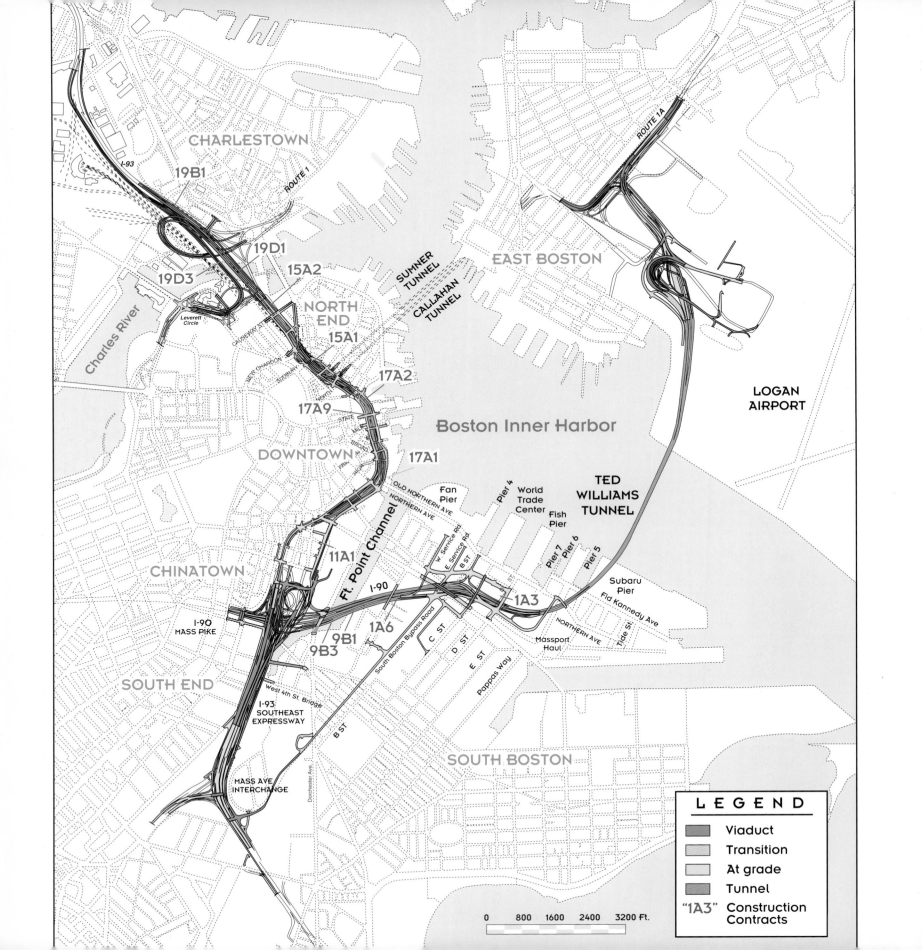

CHARLESTOWN

I-93

ROUTE 1

19B1

19D1

19D3

15A2

Leverett
Circle

Charles River

NORTH
END

15A1

NEW CHARDON

SUDBURY

CAUSEWAY ST.

17A2

17A9

NORTH

STATE

MILK

BROAD

FEDERAL

PURCHASE

DOWNTOWN

17A1

SUMNER
TUNNEL

CALLAHAN
TUNNEL

EAST BOSTON

ROUTE 1A

LOGAN
AIRPORT

Boston Inner Harbor

Fan
Pier

OLD NORTHERN AVE.

NORTHERN AVE.

Pier 4

World
Trade
Center

Fish
Pier

Pier 7
Pier 6

Pier 5

TED
WILLIAMS
TUNNEL

Subaru
Pier

Fid Kennedy Ave.

NORTHERN AVE.

Tide St.

11A1

Ft. Point Channel

I-90

CHINATOWN

I-90
MASS PIKE

9B1

9B3

1A6

W. Service Rd.

E. Service Rd.

B ST.

A ST.

1A3

C ST.

D ST.

E ST.

Pappas Way

Massport
Haul

SOUTH BOSTON

SOUTH END

West 4th St. Bridge

I-93
SOUTHEAST
EXPRESSWAY

B ST.

Dorchester Ave.

South Boston Bypass Road

MASS AVE
INTERCHANGE

LEGEND

Viaduct

Transition

At grade

Tunnel

"1A3" Construction
Contracts

0   800   1600   2400   3200 Ft.

The Big Dig will at long last complete President Eisenhower's original interstate plan when it connects I-90's current terminus with the airport. It may sound simple, but rest assured, it's not.

The project replaces an under designed and over used 1950s-era, raised roadway. That roadway, which locals call "The Green Monster," was the first elevated highway built in Massachusetts, and suffers from all the ills of a prototype. It was built before the U.S. Interstate Act that created today's Interstate Highway System. A 40-foot high wall of green steel and concrete, it tears apart neighborhoods and divides the city from its waterfront. Because it was built before federal standards applied, it fails to meet the most basic safety requirements of the interstate system. Breakdown lanes do not exist, curves in the road are too severe, and far too many off- and on-ramps create merging hazards. Just ask any of Boston's long-suffering drivers what it's like on the "Artery."

When it is complete, the Big Dig will have been a 30-year undertaking, requiring billions of dollars, untold hours of labor, and thousands of pieces of complex equipment. The most advanced technology and highly skilled people in the world have come together to execute this undertaking. Each day during its peak construction years, the Big Dig has consumed the full-time labor of nearly 5,000 employees, roughly 4,000 of whom work in organized trade unions. It is estimated that for every person directly employed on the Big Dig,

A welder repairs a drilling rig during the construction of the Ted Williams Tunnel.

two others are indirectly employed on a full-time basis as support. The total network of people is 15,000 strong.

During the conceptual phases of the project, planners recognized that the construction of an underground highway system through the heart of a 370-year-old city would be disruptive, to say the least. They pledged to minimize that impact as far as humanly possible, and even though drivers and pedestrians have been baffled, and often angered by all the construction, the planners have basically succeeded in keeping their pledge. While the work goes on, the city remains open for business, accessible to citizens, corporations, visitors, and tourists. Of the projected $15,000,000,000 budget, nearly $5,000,000,000, or one-third of the total costs, have been set aside to keep interstates, streets, and sidewalks open. Police details, temporary military bridges, modular, and reusable sections of highway are deployed throughout the city. Overnight, road crews redirect parts of the interstates, local roads, and sidewalks so that the daily work of Boston can continue as construction proceeds. Workers build tunnels underground, while above them Boston's steady roar of traffic hurtles on. Through it all, Bostonians have grown accustomed to frequent changes in their commutes. The road may change but it remains open.

Before the project is finished, a 100-acre city dump, 44 acres of an industrial corridor, and 27 acres of prime, downtown real estate under the decrepit, elevated highway will be converted from urban blight to desirable open spaces and parks. The Big Dig is also rebuilding Boston's dilapidated infrastructure. It has already improved the city's 150-year-old utilities network and it continues to fund improvements in its public transportation network. New tunnels, bridges, and roadways already opened have improved traffic conditions. Open spaces in the heart of downtown will soon be created to provide acres of parks that will enhance Boston's status as a world-class city.

Not bad for "a little road project."

Top:
An aerial photo in downtown of the elevated highway as it passes between the Boston Harbor Hotel (right) and International Place (left) in the summer of 1998. Construction activity can be seen on either side of the six lanes of elevated highway.

Bottom:
The dimensions of the future parks and surface road that will cover the eight to ten lane superhighway tunnel.

CHAPTER TWO

Boston's Dirty History

WE ARE NOT BUILDING A TUNNEL; we are constructing an underground superhighway." That's the way one engineer for the Big Dig puts it. Cutting Boston open and forcing 42 miles of underground highway in a path over 200 feet wide is one of the great engineering challenges of our time. Especially when the engineers must navigate past archaeological artifacts, old docks, glacial boulders, sunken ships, live utilities, hazardous waste, building foundations, slippery clay, and solid bedrock (the earth's crust). Imagine the Panama Canal being built through New York City and you have a pretty fair picture of the project.

Boston has been digging into its dirt for a long time. Like most early settlements in America, it was founded by emigrants from England searching for a

The underbelly of Atlantic Avenue. Gas, electric, water, sewer, thermal and telephone lines hang below the temporary roadway as construction takes place. Every week, millions of commuters, tourists and residents pass over this maze of utilities with no idea of the activity below.

place to practice their religion away from the watchful eye and oppressive hand of the English monarchy. But the early Bostonians were as industrious as they were religious. John Winthrop's eleven ships and one thousand or so followers sailed into what would become Boston Harbor in 1630. Not long after that the new Bostonians went to work in this wilderness by the sea, building wharves, leveling hills, filling in coves, and creating land to make their new home bigger and better.

Boston was founded as an anchorage, or protected port. Early European inhabitants coveted its access to the Grand Banks, whose waters were filled with haddock and codfish. But Boston had other assets, among them a deep-water port, one with a competitive advantage over the other major deep-water ports—New York and Philadelphia. Boston was the closest of the three to London's profitable trade and markets, a fact the early settlers were quick to exploit. In other words, Bostonians first looked toward the sea for everything they needed and built their city in that direction. They gave little attention to expanding their development inland or along the city's natural shorelines. Smaller settlements to the north and south—Charlestown, Dorchester—complicated expansion in those directions. It was easier to create real estate by building long wharves out to sea than to annex a neighboring village.

Topographically, few cities in the world have undergone the changes Boston has. Of the city's land, 70 percent is man-made. Almost all of the Big Dig's digging and building takes place on this invented material, or "historical fill" as it is referred to on the project. Historical fill is just about anything the early Bostonians could grab hold of and throw into the harbor as landfill. This weak, wet, unpredictable material is possibly the most difficult substance to build a superhighway tunnel through.

In 1643 the first "landmaking" venture was completed at the north cove, near today's North End. The partnership of George Burden, John Button, John

Boston has. Of the city's land, 70% is man-made.

Hill, and Henry Symons dammed the harbor's north cove and made Mill Pond Boston's first Dig. The agreement between the town and the partnership specified that the partnership "dig trenches in the highways . . . and make and mainteyne sufficient, passable and safe wayes over the same for horse and Cart." The Dig converted the north cove into a millpond that supported a grist-mill, a sawmill, and a chocolate mill. Next the joint venture exercised its right to dig through their millpond and connect to the harbor, creating Mill Creek. This same creek was recently uncovered during tunnel construction, along with two enormous, well-preserved 2200-pound millstones.

BUILDING AN UNDERGROUND SUPERHIGHWAY THROUGH HISTORY

To understand what the Big Dig engineers and construction managers are dealing with, let's take a quick trip through Boston's topographical history.

The early settlers of Boston soon began their wharfing-out projects. One Captain Oliver Noyes and his crew gathered together masses of timber, rocks and dirt and started to pile them from the end of King Street, now State Street, out into the town's harbor. Their structure enabled the largest ships of the day to load and unload their goods, and supported the shops and warehouses that sprang up along its sides. When Noyes and his crew were finished, they'd built an 800-foot long avenue extending out to the sea—a lifeline connecting the New World to the old country. The structure has been incorporated into Boston's longest wharf, known, not surprisingly, as Long Wharf.

By 1742, 166 wharves and docks had been constructed along Boston's waterfront. They were set out like spokes of a wheel extending from the hub. As ships from the 1600s became the wider and longer ships of the 1700s and the even larger and deeper-hulled steamships of the 1800s, the wharves were extended farther and farther into the harbor's deeper waters. Problems developed when these protrusions collided with the flow of water from the Charles

Opposite:
In 1630, Boston was only 30% of what it is today. During high tide, Boston became an island as the peninsula's sandy neck was covered with water. Over the years, aggressive landfill operations filled in the city's shorelines and added 70% to the small island.

and Mystic rivers into Boston Harbor. As the moving river water ran into the long wharves and docks of the harbor, sandy sediments, debris, and rubbish settled in between the wharves. The debris blocked the access of the larger ships, and the pier owners were forced to extend their structures farther into the harbor. This cycle continued for centuries.

To cover the cost of building, wharf owners, making the best of the situation, created land between the wharves and then developed and sold this newly made property. The spaces between wharves were filled with mud, timbers, gravel from the tops of local hills, and even rubble from the Great Fire of 1872 that destroyed 65 acres of Boston's business district. One of the more ingenious methods of landmaking was filling the hull of a junked ship with rocks, scuttling it at dockside, and filling earth in and around the hull. Big Dig construction crews have come across remains of these buried ships well into the downtown, hundreds of feet from the water.

FIRSTS AND LARGESTS

Boston has a rich history of firsts and largests in the nation and the world. Alexander Graham Bell's famous first phone call and King Gillette's safety razors made their world debuts in Boston. The first World Series baseball game was won in Boston by the Boston Pilgrims in 1903, and the first organized soccer game in America was played on Boston Common. The first major public library system, and the first mass transit system in the United States began in Boston. Harvard, in nearby Cambridge, is America's first college. And now the country's largest construction project—the Big Dig—is building its way through the city.

Boston also claims bragging rights to the first subway system in the country and the largest train station in the world. The first subway car ran on September 1, 1897, from Park Street to the Public Garden. London, Budapest,

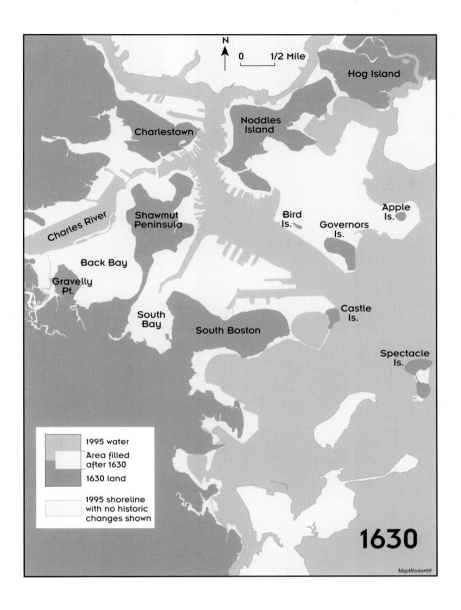

1630

Legend:
- 1995 water
- Area filled after 1630
- 1630 land
- 1995 shoreline with no historic changes shown

MapWorks•99

Labels on map: Hog Island, Noddles Island, Charlestown, Charles River, Shawmut Peninsula, Bird Is., Governors Is., Apple Is., Back Bay, Gravelly Pt., South Bay, South Boston, Castle Is., Spectacle Is.

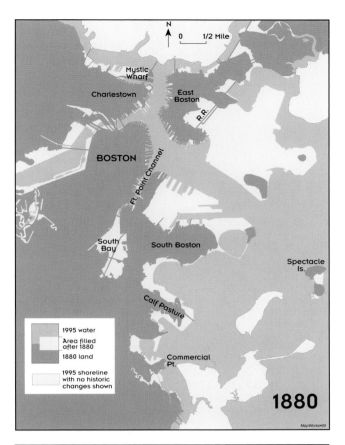

1880

Legend:
- 1995 water
- Area filled after 1880
- 1880 land
- 1995 shoreline with no historic changes shown

MapWorks•99

Labels on map: Mystic Wharf, Charlestown, East Boston, R.R., BOSTON, Ft. Point Channel, South Bay, South Boston, Spectacle Is., Calf Pasture, Commercial Pt.

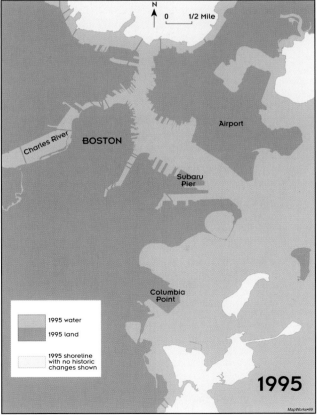

1995

Legend:
- 1995 water
- 1995 land
- 1995 shoreline with no historic changes shown

MapWorks•99

Labels on map: Charles River, BOSTON, Airport, Subaru Pier, Columbia Point

and Glasgow were the only other cities in the world with subway systems at the time. In 1898 Boston's South Station opened its gargantuan train shed and became the biggest train facility ever, with hundreds of trains arriving and departing each day. In 1904 the aggressive engineering of the Boston Transit Authority produced America's first underwater tunnel to carry a train—the tunnel that evolved into the city's Blue Line.

Opposite:
Starting from the
north and working
south, Commissioner
William Callahan
began construction
of the Central Artery
in January of 1951.
Thirty-six acres of
Boston's downtown
were destroyed
(Boston's North End
is to the right).

The Big Dig will intersect, bob, weave, and duck its way around the existing Blue Line subway tunnel, built in 1904, the Red Line subway tunnel finished in 1917, and the Orange Line tunnel constructed in 1975. The subterranean superhighway is being built directly over the Blue Line and under and over the Red Line.

Gridlock, congestion, and traffic have been Boston's curse from its earliest days. By the 1880s the heart of Boston had become so congested with streetcars that it was said that a passenger could reach his destination faster by climbing onto the roof of his trolley and walking over the tops of stalled vehicles. In 1892, the city surveyor's report declared "no scheme of street improvements which does not increase the sum of north and south widths . . . can hope to successfully cope with this stupendous and growing problem of congestion in downtown Boston." The crowded, noisy, polluted transportation corridor that Interstate 93 now occupies has been Boston's bane for well over 100 years.

Boston's urban planners turned to the idea of an elevated highway as early as 1930. The city's flamboyant mayor, later governor of Massachusetts, James Michael Curley—he was elected mayor while indicted on mail fraud charges—was a moving force behind the idea. But political rivals prevented him from building it. In the early 1940s Governor Leverett Saltonstall and Mayor Maurice Tobin rallied the business community to build the elevated highway, but World War II brought the idea to an abrupt halt. Finally, in 1949, William Callahan, commissioner of Public Works, made the construction of the elevated highway a top priority. Callahan was known as a tough bureaucrat, willing to go to great lengths to get his way. To some he's Boston's version of Robert Moses, New York City's autocratic master planner.

On January 22, 1951, construction began at the Charles River on what would become an eight-year ordeal of destruction and upheaval for some of

Boston's poorest neighborhoods. The new highway was slated to begin in the North End—a close-knit Italian community of old brick buildings, winding streets, and mom and pop businesses. From there it would proceed south through Chinatown—another close-knit community—and the small leather and wool factories near South Station. It's hard to think of a path that could cause more upheaval to more people than the one the planners came up with.

The destruction was overwhelming:

Opposite:
The original highway was 3.7 miles long and was finished in 1959. It created a steel wall between the city and the water. Today, the Central Artery is one of the most crowded roadways in the country. It was designed to handle 75,000 vehicles per day and now carries nearly 200,000.

1000 residential and commercial structures were demolished, displacing over 20,000 residences and businesses. In the North End, families and business owners fought to have the highway moved toward the waterfront and away from their neighborhood, but Callahan could not be swayed. Boston's downtown, like the cores of many American cities, was in a steady decline. Easing access to the city's business and retail center was far more important to Callahan than the wishes of a small, poor immigrant enclave.

As the Public Works' crews bulldozed their way through the city, opposition to the highway's construction grew. The noise, dirt, and blight in the North End were horrible, and neighborhoods in Callahan's path, anticipating the same fate, began to organize and fight. Even those who had originally supported the project began to have doubts.

Thomas "Tip" O'Neill, Jr., who was speaker of the Massachusetts House of Representatives when funding for the road was approved, recalled that, "Everything was done by the greatest engineers and architects in the country. You took for granted that their advice on road location was correct."

But it wasn't. The elevated highway, infelicitously painted a dull green, became known as the Green Monster. A hulking 40 feet high, 200 feet wide, and 3.7 miles long, it became a steel wall, cutting off Boston from its historical waterfront and harbor.

In 1952 a newly elected governor, Christian Herter, named John Volpe Public Works commissioner, replacing Callahan. Volpe, who would later become U.S. Secretary of Transportation, was said to have had a window overlooking the elevated highway during its construction and promised he "would never build anything that ugly." One of his first acts as commissioner was to stop construction on the Central Artery, which had already plowed a mile-and-a-half-long path through Boston's downtown. Next Volpe planned and constructed the widest known vehicle tunnel in the world at that time.

This tunnel, now called the South Station or Dewey Square Tunnel, was Volpe's answer to Callahan's mistake. Soon the more than 2000 foot tunnel became an inspiration as future politicians struggled with Boston's worsening traffic problems and ugly highway.

Boston's $110,000,000 Highway in the Skies

By K. S. BARTLETT

92,000 Tons of Steel.

How soon can we use it?

construction of the towering steel skeleton of the elevated half of the highway has been watched by scores

CHAPTER THREE

All Politics Are Local

IGNIFIED RESTRAINT has never been a hallmark of Boston politics. And it was the lack of restraining federal guidelines that allowed Callahan's elevated highway to evolve into the Green Monster.

Since the highway was begun before the law that created the Interstate System was enacted, it was built exclusively with city and state funds—and the state built it cheap and to its own specifications. Today that section of I-93 falls short of the most minimal requirements of an interstate highway. Its curves are too sharp, the lanes too narrow, and too many off- and on-ramps clog the system. There are neither acceleration lanes for cars preparing to make high-speed merges nor deceleration lanes for drivers slowing down to exit the system.

During peak construction years, the Big Dig spends nearly $110,000,000 a month, which was the cost of the original Central Artery highway built through Boston's downtown in the 1950s. The "highway in the skies" was an eight-year project built with heated off-ramps that never worked and were abandoned.

There are no breakdown lanes and sections of the highway are so grossly inadequate that the familiar red, white, and blue Interstate signs are not posted.

From the day it opened in 1959, the elevated highway's 34 off- and on-ramps, totaling 4.3 miles, choked the 3.7 miles of roadway. Think of it: more ramps than highway. That's proven to be a lethal combination, resulting in more congestion than on any other highway in the country—between 25,000 and 30,000 vehicles per lane a day at its worst choke points. That is more traffic than New York City's West Side Highway was carrying before its elevated sections buckled and split apart in the early 1970s. The accident rate for the elevated highway is four times the national average for urban interstates.

Debates about widening the Central Artery began as soon as construction was completed. There were even discussions of adding another level to the top of the already elevated highway, a level that would have doubled its height to almost 80 feet! But the only realistic way to widen or increase its capacity was to tear down more of the neighborhoods that had already been battered by the wrecking ball. And by now the affected neighborhoods were fighting back, loudly resisting any such plans.

What to do?

The answer ironically was with the Feds. Federal funds for highways became available in 1956. With access to these funds, Massachusetts, along with many other states in the country, began what an independent transportation study later called a "pathological" campaign of road building—a "great mindless system charging ahead." The country was building roads not because they were needed but because the federal government was paying 90 percent of the cost. The states contributed only ten cents for each dollar spent on building the interstate system. With "ten-cent dollars" from Washington pouring into the states' coffers and a requirement to use them or lose them, any governor not building more highways was putting his political future at risk.

Frank Sargent, governor of Massachusetts from 1969 to 1974, took that risk. With strong anti-highway and pro-mass transit groups battling entrenched pro-highway groups, Sargent found the middle ground. In 1970 he stopped the extension of I-95, known also as the Inner Belt, from demolishing 3800 homes as it passed through the city of Boston. Sargent, in fact, halted work on all major highway projects and began a study of alternatives. As he explained the rationale for this abrupt action, "Nearly everyone was sure that highways were the only answer to transportation problems for years to come. But we were wrong." Sargent, along with a group of governors from around the country, led a successful effort to allow states to exchange federal highway funding for mass transit projects. The resulting legislation made federal "highway" funds available for improvement of public transit systems. This flexibility turned out to be critical in the case of the Big Dig, because Boston's mass transit system, the T, would need substantial and immediate improvements if it was to support heavier commuter use during the years of the Big Dig highway construction.

Enter Governor Michael Dukakis. When Dukakis defeated Governor Sargent in 1974, he completed a pendulum swing from the pro-highway building 1950s and 1960s to the environmentally conscious, pro-mass transit 1970s. Governor Dukakis, it was said, regularly rode Boston's Green Line subway to his office at the State House, on Beacon Hill.

Governor Michael Dukakis was considered pro-mass transit and anti-highway. In a classic Big Dig paradox, he launched the biggest highway project in American history. And he says he would do it all over again.

He was so anti-highway at the time that former state senate president Kevin Harrington, who battled Dukakis on these issues, remembers, "To Dukakis, highways and automobiles were . . . animate objects which he hated as you would hate individual evil in people. . . . It's very hard for people today to understand almost the fanaticism he had about transportation—and I am choosing my words carefully. . . . Nothing ruffled Mike Dukakis except transportation. I've often wondered if you could get him on the psychiatrist's couch and ask, 'What was it that really stirred the passions like that?'"

Ironically, nearly ten years later, in 1983, Governor Dukakis would preside over Massachusetts as it prepared for the Big Dig, combining two large highway projects into the biggest public works project in U.S. history. The first project, the depression of the elevated highway through downtown, had Dukakis's support from the beginning. He originally opposed the second project, the construction of a third vehicle tunnel under Boston Harbor, which was supported by his political rival, Ed King. (Two other tunnels—the Sumner Tunnel and the Callahan Tunnel already existed.) King, who ran against and defeated Governor Dukakis in 1978, was backed by a pro-highway, pro-business coalition that wanted another tunnel under the harbor to service Boston's Logan International Airport.

Dukakis, in fact, was so opposed to a third tunnel that he made it a major issue in his campaign to regain office. Running against Governor King again in 1982, candidate Dukakis told a cheering crowd of 300 East Boston supporters, "I don't know why in 1982 we're talking about spending half a billion for another tunnel. There's no reason, no excuse and no need for it."

That was just before Fred Salvucci, Dukakis's quiet, unassuming but take-no-prisoners Secretary of Transportation, saw the benefit of building both projects—together.

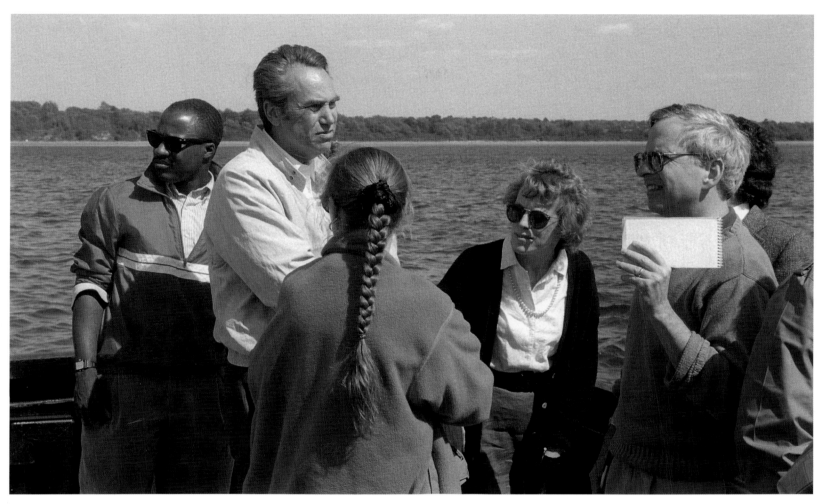

Fred Salvucci
(second from left)
talks with Laura Brown
from the *Boston Herald*
(back to camera) and
others on a boat trip
to Spectacle Island
in 1989. Without
Salvucci's vision and
political tenacity the
Big Dig would never
have been possible.

THE MAN WHO CREATED THE BIG DIG

Frederick P. Salvucci is a cross between an Albert Einstein of transportation planning and television's Mr. Rogers. He is soft-spoken, mild-mannered, and listens to groups more than he speaks to them. He is a visionary who, once committed to a cause, never gives up. As a result he has spent nearly his whole public life pursuing the depression of the Central Artery and reuniting the city with its waterfront.

Former Boston mayor Kevin White, who knew Salvucci as a young man and was one of the first to recognize his talents, says it best: "Fred has two enormous strengths, when few of us have one. First, he is technically proficient in the world of transportation. Second, he is a politician to his fingertips, in the best sense of the word. . . . He does not give away power, he aggregates it and when necessary, he delegates it . . . he is focused." Fred Salvucci is the man behind the Big Dig. Without his leadership and political skills, this amazing project would never have seen the light of day.

Fred Salvucci grew up in Boston and attended the Massachusetts Institute of Technology, where he received his master's degree in civil engineering. After graduating, Salvucci became engrossed in Boston's transportation issues and planning. Living in Boston's North End and working for the newly created Boston Redevelopment Authority, Salvucci was part of a group studying plans to widen the elevated highway and add more ramps to alleviate congestion. Even as a young urban planner Salvucci opposed these ideas, partly because he thought them ineffective and partly because they required more demolition of the North End.

Salvucci had witnessed the hostile eviction of his grandmother from her home in the Brighton area, by the extension of the I-90 Massachusetts Turnpike in the early 1960s. Salvucci's grandmother and her neighbors were issued a one-dollar down payment for their homes and given thirty-day eviction notices. Everything they owned—furniture, family mementos, dogs, cats—had to be out of their homes and relocated. A month later the homes were demolished. The dispossessed homeowners received only 50 percent of the appraised value as a final payment.

At that time the head of the Turnpike Authority was none other than William Callahan, the same man who led the Department of Public Works in 1951 when it began destroying Boston neighborhoods to build the elevated Central Artery. Instead of following Callahan's example, Salvucci pursued improvements to pedestrian routes around the Central Artery, arguing that congestion on the elevated highway was caused by a little old lady crossing the street in the North End, not by a lack of highway ramps.

Salvucci's reasoned arguments caught the attention of Kevin White, then mayor of Boston, who put him to work as his transportation consultant. The

hate roads and it's because the Central Artery is so congested."

two men were perfect allies; they both hated the elevated highway. Salvucci remembers Mayor White gazing out his city hall office window, looking onto the elevated highway, and saying, "You goddamn engineers, you cut my city in half with that thing." Salvucci was determined to heal the "ugly green scar" across the downtown, but just how to do it was still a mystery. The final solution did not come to light until Bill Reynolds, another M.I.T. trained engineer, called on Salvucci to discuss the issue in the early 1970s.

Bill Reynolds, a contractor who repaired and expanded highways in and around Boston, often debated with Salvucci on transportation policy. Salvucci recalls first discussing the Big Dig at Jacob Wirth's restaurant over coffee and sandwiches. "One day Bill said to me, 'Highways are beautiful things, they are part of the American way of life, but in Boston, people hate roads and I've become convinced it's because the elevated Central Artery is so congested. It's like a giant neon sign flashing and saying, "Roads are bad, roads are bad." It's horrible to drive on. It's ugly to look at. So I have concluded that the only way we are going to change the anti-highway bias in Massachusetts is to put it underground.' "

Salvucci at first rejected the notion, saying, "Gee, Bill, three quarters of the city already thinks I am crazy. Why should I convince the other quarter?" But then, after studying the idea as an engineer, he became fascinated with the concept. Today he admits to the *Metrowest Daily News*, "without Bill it never would have started." Reynolds in return gives Salvucci credit for getting the project approved and making his idea work.

Salvucci, as it happens, has a habit of spending his lunchtime wandering around Boston. One day he was walking under the elevated Central Artery, and as he walked, he mentally measured the space around and under the existing structure. Sure enough, there was sufficient room to build an eight-lane tunnel.

"Reynolds's idea wouldn't get out of my head," he remembers. "I started to look at the amount of space between the ramps. I realized there's actually enough room there, you might be able to sequence the construction so you don't shut the city down."

He became convinced that the existing Central Artery could be supported while a tunnel was built beneath it. And that he could do it with minimal disruption to city traffic. Salvucci gained the support of his boss, Mayor White, of Governor Sargent and perhaps most importantly, of *The Boston Globe*, Boston's largest newspaper. In fact, Mayor White and Governor Sargent paid Salvucci the highest possible compliment and strongest endorsement a politician can offer—they each claimed the idea was his own.

Bill Reynolds had hoped that the Central Artery would come down sooner rather than later. He recalls, "I figured, take it down and get it built for the Bicentennial in 1976, and we would all be happy. That's the way jobs used to get done. We weren't talking billions. I never really was thinking of the costs, they were just normal costs."

When Dukakis first won the governorship in 1974, he chose Salvucci as Secretary of Transportation and supported his efforts to depress the Central Artery. Fred Salvucci was becoming defined by his vision to bury the elevated highway. However, all plans went on hold when Dukakis lost his reelection bid in 1978 to King. Salvucci stepped down as Secretary of Transportation and returned to M.I.T. to teach engineering. But he hadn't yet retired from the fray.

Governor Edward J. King held office for only one term, from 1978 to 1982. He ignored Salvucci's plans for depressing the Artery and concentrated on building a third tunnel between the downtown and Boston's Logan airport, cutting through East Boston, another densely settled neighborhood. Like the earlier projects, King's third harbor tunnel required the taking of homes and businesses for tollbooths and feeder roadways. King had the support of many businesses, which wanted improved access to the airport, and of the construction industry, which wanted to build the tunnel. But he lacked the broader political support he needed. East Boston had already been torn up with the building of the two other tunnels and the constantly expanding Logan airport. The community strongly opposed Governor King's plans, fought back furiously, and prevented the tunnel from being built under his watch.

> *"The Big Dig is a big deal. Its scope and scale are daunting. But it's like workers underfoot in your house. You want them done and you want them out already."*
> —Sally Abrahms, Brookline, MA

While all this was going on, Bill Reynolds was still mulling over his idea of depressing the Central Artery. One day he placed a call to Fred Salvucci from a pay phone near Boston's South Station. Salvucci was still at M.I.T., and planning Michael Dukakis's gubernatorial comeback. This time Reynolds was calling about the third harbor tunnel that King was trying to build.

Salvucci remembers, "Bill Reynolds called me and said, 'What are you doing?' I said, 'I am correcting papers.' He said, 'I have a little job doing some paving and track laying for Amtrak at South Station and I can see an alignment for the third tunnel. You have to come and look at it.' I said, 'Why are you talking to me? Call the Secretary of Transportation.' He said, 'They're crazy. They won't listen to me.'

"So we drive over and we take a look. You can physically see it because that section of South Boston is basically unutilized parking lots and rail yards. It is a vacant swath right across South Boston. What he said was, 'This crazy plan of King's isn't going to work. If you go through East Boston, Tip O'Neill [Speaker of the U.S. House of Representatives, who represented East Boston] is going to stop it. If you swing over to the airport, the curve is so violent that Federal Highway is not going to like it. The right way to do it is to come across South Boston and over to the airport.' I go over, look at it, and said, 'Gee, you are right.'. . . This was really brilliant. He had solved the whole problem by changing the alignment."

In 1982 Michael Dukakis defeated Governor King and was elected to his second term in office. He soon reappointed Fred Salvucci as Secretary of Transportation. Even while out of office, Salvucci had quietly begun to seek support for an ambitious combined campaign, the depressing and widening of the Central Artery and the building of a third harbor tunnel to the airport—without destroying neighborhoods. In 1983 *The Boston Globe* ran a story about Secretary Salvucci's plans. Dukakis, surprised by the story, was angry and demanded to see "Governor Salvucci." Alerted by a kind word of warning from the governor's secretary, Salvucci approached the Governor's office with care. As Salvucci explains, "I worked every bank shot I could to get Mike to feel comfortable that this was the right thing to do both in the terms of politics and on the merits."

Salvucci won over his boss and so paved the way for the combined project. This would be the first highway project in decades in Massachusetts to win broad support and face little opposition. On the city, state, and regional levels it was understood that even during peak construction, the Big Dig would allow businesses to remain open. With luck, almost all the money needed would come from the Federal Highway Administration. The public, nonetheless, realized it was an expensive undertaking. State Legislator Barney Frank joked about the costs saying, "Wouldn't it be cheaper to raise the city than to depress the Artery?"

Now on to Washington, D.C., for the federal money. No problem. The powerful Thomas "Tip" O'Neill, Jr., was on their team, and the Massachusetts delegation was confident their project would be funded. As Salvucci explained in a 1998 History Channel interview at M.I.T., there was one thing they didn't know. President Ronald Reagan was going to pick a fight with that "pack of communists" in Massachusetts.

TIP'S GOING-AWAY PRESENT

House Speaker Tip O'Neill was famous for saying, in his inimitable way, "All politics is local." It was the cornerstone of his political dynasty, and as he explained in an interview just prior to his death, "I had opposed the Third Harbor Tunnel . . . because they gave us a choice, putting the Central Artery under or building the Third Harbor Tunnel. I was insisting on the [Central Artery] being put down below. When Jimmy Howard came to me and said, 'Hey, we can authorize both,' there was no problem. . . . My only question was 'What was going to be the effect in East Boston?' which I represented. When they decided . . . no homes were going to be destroyed, no areas of playgrounds were going to be taken, [and] nobody in East Boston was going to be hurt economically, I didn't have any difficulty with it."

With local backing in Boston secured, Massachusetts officials thought authorization on Capitol Hill would be a mere formality. In 1984 Tip O'Neill announced that he was stepping down as Speaker of the House, effective in 1986, after a long and legendary career. House Public Works chairman James Howard, an old friend of O'Neill's, and other colleagues in the House began to talk about "Tip's Tunnel" and "Tip's going-away present." No one expected the issue of authorization to drag on beyond O'Neill's retirement. But it did.

The problem was nothing less than the opposition of the President of the United States. Ronald Reagan's White House fought the Central Artery/Tunnel Project every step of the way. Reagan's Secretary of Transportation, Elizabeth

Dole, was adamantly opposed to funding the Massachusetts project, claiming that the cost "is not justified on the basis of the transportation benefits to the nation." Secretary Dole's federal highway administrator, Ray Barnhart, a loyal "Reagan Republican" from Texas, bluntly recalls, "Massachusetts had screwed around for so many years in such an anti-transportation mode [anti-highway] with all their social maneuvering. Coming from Texas, my first inclination was to let the bastards freeze in the dark." The environment was hostile, and soon the national media picked up on the battle between two political titans, Speaker Tip O'Neill and President Ronald Reagan. The Central Artery/Tunnel Project was already a national debate, and not one spade of earth had been turned.

THE BIG DIG WINS BY ONE VOTE

In 1986, House Speaker Tip O'Neill stepped down as promised. The new House Speaker, James Wright of Texas, pushed a transportation bill, including Big Dig funding, through the House. That bill, the Surface Transportation and Uniform Relocation Assistance Act of 1987, also passed through the newly elected Democratic Senate with overwhelming support. President Reagan vetoed the bill on March 27, 1987, attacking the largess of the Central Artery/Tunnel Project. *The Washington Post* reported Reagan as saying, "I have not seen this much lard since I handed out blue ribbons at the Iowa State Fair."

The House overrode President Reagan's veto on March 31, but the Senate, voting on April Fools' Day, 1987, sustained the veto. A bit prematurely, the White House began to celebrate its veto victory. Vigorously working their colleagues, Massachusetts Senator Edward Kennedy, allied with West Virginia Senator Robert Byrd, obtained reconsideration. A re-vote was in the works.

The strategy was simple and blunt. Senator Kennedy and others threatened to pull tobacco subsidies from North Carolina if Senator Terry Sanford did not change his vote and support the bill. When word of the new maneuvers drifted back to the White House, President Reagan himself made a special trip to Capitol Hill in a last-ditch effort to secure the veto. But Senator Sanford had been surrounded by backers of the bill, and on April 2, cast the critical vote overriding the veto. That vote was extracted from him by classic strong-arm, power politics. A member of Senator Robert Dole's staff, Jim Whittinghill, remembers, "If you watch closely, you will see switches around here every so often. But you hardly ever see something like that. It was brutal."

The transportation bill became law in April of 1987, and the Central Artery/Tunnel Project had its federal funding. The Big Dig was now a reality. Back home, the *Boston Herald* ran a headline saying, "Relief! Artery Dream is now reality."

CHAPTER FOUR

Gearing Up

N 1985, TWO YEARS BEFORE the money was approved in Washington, D.C., Fred Salvucci was already searching for the strongest team of engineers and managers in the world. The Big Dig was gearing up, and Salvucci knew that getting the project through the permitting, design, and construction phases would be difficult.

After assessing the demands of the Big Dig against the available skills, talents, and construction experiences of the various bureaucracies in the Commonwealth, Salvucci decided to use both public and private manpower. The Massachusetts Department of Public Works would take the lead role in overseeing a private management team. Together they would coordinate, direct, schedule, and review design and construction activities on the Big Dig.

Five groups with highly technical experience in

"The Big Dig is the most heavily permitted project in the history of the world!" Former project director Peter M. Zuk often made this claim as he guided the project through the difficult stages of approval. Shown left are volumes of the Environmental Impact Statement.

subterranean construction competed for the first four-month, $1,300,000 consulting contract. In 1985 a committee of nine highway officials unanimously selected the joint venture of Bechtel Corporation of San Francisco and Parsons, Brinckerhoff, Quade & Douglas of New York, two of the largest consulting and management-engineering firms in the world. Bechtel/Parsons Brinckerhoff, or the Joint Venture as it is known today, has been the management consultant of the Big Dig since 1985 with staffing levels as high as 1000.

The vote to select the two firms was unanimous because of their combined skill in slurry-wall construction (the construction of underground walls). The Big Dig requires the largest use of slurry walls on one project, ever. As Fred Salvucci observed, "Between them, Bechtel and Parsons Brinckerhoff had done 90 percent of the slurry-wall construction in the United States. If either had joined another team, there might have been a choice. By going together, they blew everyone away."

THE JOINT VENTURE

When the Joint Venture began work on the Big Dig in 1985, Bechtel was the largest construction firm in the world, and Parsons Brinckerhoff was the seventeenth largest design firm in the world, according to *Engineering News-Record*. They had previously worked as a joint venture on San Francisco's Bay Area Rapid Transit (BART), the first new transit system in the United States since World War II, and the Metropolitan Atlanta Rapid Transit Authority (MARTA).

Both had worldwide reputations for excellence. Bechtel was part of the team that built the Hoover Dam and, in 1951 developed the world's first experimental nuclear power plant to produce electricity. In 1962 the Bechtel family built Canada's first nuclear plant and the company was heavily involved in building the Eurotunnel's English Channel tunnel-rail link project, the "Chunnel."

Parsons, Brinckerhoff, Quade & Douglas engineered and designed the

Rogers Pass railroad tunnel in Canada—the longest tunnel in the Western Hemisphere—and Baltimore's Fort McHenry Tunnel—the widest underwater tunnel in the world—two projects that positioned them well for work on the Big Dig's tunnels.

FROM DAVID TO GOLIATH

A funny thing happened to the Big Dig on its way to becoming a reality. It had begun as a long shot in a tough political battle. By the time the proposal left the State House in the summer of 1983 and traveled to Capitol Hill in Washington, D.C., its chances were good; but success was far from certain. When it arrived back in Boston in the spring of 1987, four years later, the project was no longer the underdog. It was now "the fat kid on the block with a few billion in his pocket," according to Fred Salvucci. David had become Goliath. "Paradoxically," Salvucci continues, "the politics got even tougher. Special interest groups, government organizations and individual communities all wanted a piece of the well-funded action. The problem was that for every month that construction was delayed by partisan demands, the price tag rose eighteen million to cover inflation."

After receiving authorization from the U.S. government in 1987, Salvucci expected to spend six months to a year resolving environmental issues and obtaining permits for construction. Three-and-a-half years later the federal government finally gave the project partial environmental approval—to begin construction south of the Charles River. The river itself and the area north of it were still a problem.

To proceed, even partially, the project had to get past two large environmental hurdles. First, it needed to secure approval of its Environmental Impact Report (EIR) from the Massachusetts Executive Office of Environmental Affairs. Then, after the commonwealth approved the EIR, the federal government had

Highly contaminated materials on Spectacle Island, the Fort Point Channel, and the bottom of Boston Harbor required special handling and protection gear for workers.

THE BIG DIG · Gearing Up

to give its own approval in a process involving reams of documents known as the federal Environmental Impact Statement (EIS). Both the EIR and EIS are made up of volumes of requirements, rules, and obligations that the project is held to legally. For example, over 1500 commitments to the communities were listed in the final EIS, ranging from rat control to street sweeping.

Myriad diverse groups—some with conflicting agendas—had to agree if the Big Dig was to secure the approvals. Fearing a breakup of the fragile alliance that supported the project, Salvucci allowed neighborhoods, business-es, environmental groups, and other bureaucracies to modify the project's final design in exchange for their backing. Every time the project's scope or design changed, the entire plan was re-evaluated by both state and federal agencies against the latest environmental regulations. And the regulations got tougher with every year that elapsed. Over 100 changes in the project, in addition to stricter environmental requirements, added years to the project's completion date before the digging even started.

The change that loomed largest was environmental approval for Scheme Z, the Charles River Crossing. Scheme Z was one of 31 designs. Salvucci felt Z was the best of the designs, but it was attacked from day one by the city of Cambridge and a neighboring community, Charlestown. In the end, Z was scrapped.

After three years and $1,400,000,000 of inflation, construction cost increases, and design payments, the Charles River Crossing was approved in June 1994. It had taken the Big Dig seven years to get its final environmental approvals from the federal government.

In only one of the project's many ironies, Massachusetts gave its approval of the EIR on January 2, 1991, and Governor Michael Dukakis's third and final administration ended its term the next day, January 3, 1991. Salvucci had no choice but to hand over the reins to a new administration and a new Secretary of Transportation. As a result, Fred Salvucci, who cleared the path for building the nation's largest public works project, was never to oversee any of the official construction activity of the Big Dig.

RATS!

Rats are amazing creatures. They are able to climb straight up a brick wall, fall off five-story buildings and land on their feet without a scratch. Their yellow teeth carry disease and bite with the force of a shark. They can swim for three days straight, jump more than three feet high, and squeeze through a hole the size of a quarter. They eat 25 percent of their body weight every day, and females can give birth to 70 little rats a year. Adult rats have attacked human babies and are vicious if cornered. Welcome to Boston politics and psychological warfare—the battle between a colorful mayor and a stubborn Secretary of Transportation.

It is no wonder Bostonians were frightened when the head of the city's Rodent Control Department, Sam Wood, claimed on a 1987 local television program that the rat population was higher than two rats per citizen in some areas of town. (The ratio is actually thought to be one rat for every 15-40 people.) Some city officials were telling the public that construction from the Big Dig was going to dislodge millions of rats from deep down under the city. Rumors of rats swimming up pipes and through toilets and into homes began to spread.

By early 1990, the rat scare had reached its height. CBS-TV, *Time* magazine, and *The Wall Street Journal* had all reported on Boston's rat population and the Big Dig. Mayor Raymond Flynn, the mastermind behind the successful rat scare, was demanding that Salvucci increase the size of the project's rodent control program. "Flynn wanted everything from Massachusetts Avenue to Boston Harbor to be controlled by the rodent control program. Massachusetts Avenue is miles from the Central Artery, and rats simply don't travel that far,"

Opposite:
"The Rat That Ate Boston," an illustration by artist Peter de Sève for an article titled "The Rats of Boston" that ran in the May 1989 issue of *Yankee* magazine.

Salvucci says. "I had already been extra generous with the program's geographic parameter, and FHWA [Federal Highway Administration] funds couldn't be used for cleaning up the city's entire rat population."

Flynn had gone to the media, so Salvucci went to the business community and said, "Hey, you guys are smarter than me. If you think the mayor getting national news organizations to heat up the rat issue is good for your business, you must know something that I don't know."

Pressure from the business community forced Flynn to tone down the rhetoric and he and Salvucci compromised on a rodent control program twice the size of the original but smaller than Flynn's grand plan. So ended the rat scare. It typified what Salvucci faced on many levels.

The project's "ratologist" for the first ten years of construction was Dr. Bruce Colvin, senior environmental biologist, in charge of the most comprehensive rodent control program for a construction project anywhere in the world. Dr. Colvin, a zoology graduate of Ohio State who later received his doctorate from Ohio's Bowling Green University, has a passion for rats. He has spent hours lying on cold, damp floors, allowing rats to climb over his body. He's allowed rats to nip and bite him. He even decorated his home with rat skulls. This man knows his enemy.

Dr. Colvin's primary goal was to educate the public on preventive measures. As he explained, "It's not construction that grows rats, it's garbage." He also surveyed the environment, baited, trapped, and exterminated the rodents in the construction zones before bulldozing began. His plan was to reduce or eliminate the rat population before the digging and so prevent rodent migration into the neighborhoods. His logic was simple. "Eliminate their food source and their habitat, and you will eliminate their population."

Dr. Colvin's job was to coordinate the efforts of five biologists, four pest control companies under subcontract with the project, and the city's 17 inspec-

rat skulls. This man knows his enemy.

tors from the Rodent Control Department. Educating the community by going door-to-door and handing out literature was the first step. Next they examined the environment and counted the rats, using cameras with infrared lenses, eyewitness sightings, quantity of rat droppings and devoured bait. Computers organized the information and applied it to a map. Hundreds of workers and millions of pounds of equipment could not begin construction until Dr. Colvin declared a work zone as "rat free."

In a *Boston Globe* interview, Dr. Colvin explained, "The peak of down-town digging is still two years away, but thanks to community awareness programs, improved trash can designs, and liberal doses of poison, Boston's rats have so far pretty much stuck to their traditional haunts." The rat scare, once a headline-making story, was finally being relegated to the back page.

"**T**HE MOST HEAVILY PERMITTED PROJECT IN THE HISTORY OF THE WORLD!**"** Concerns about rats and the environment are only part of the minefield that the Big Dig must negotiate every day. It is, says Peter M. Zuk, "the most heavily permitted project in the history of the world!" Zuk, project director from 1991 to 1999, often made this claim when he was explaining the bureaucratic calisthenics the project regularly performed for permit approvals. The project is so large that even after 1994, when both the state and federal governments had signed off, permitting was still an ongoing process. The Big Dig lives and dies by those pieces of paper. Construction cannot start until a permit has been obtained from an agency such as the United States Coast Guard, the Federal Aviation Administration, or the Army Corps of Engineers. Even their names are intimidating. Try building the largest highway project ever in an urban core and getting their approval—a monumental task.

With more permitting specialists than city rodent control officers, Jeffrey M. Paul, Chief Permit Supervisor and leader of 20 full-time permit

The project will place 3.8 million cubic yards of concrete, enough to build a sidewalk three feet wide and four inches thick from Boston to San Francisco... three times.

specialists, set out to beg, borrow, and negotiate approvals for nearly 500 very large permits over the course of ten years.

An important weapon in the permitting arsenal was General William Flynn's "critical issues list." Deputy project director Bill Flynn, a retired three-star general of the U.S. Army, relentlessly tracked thousands of permitting issues that could delay construction. Any hapless person whose name appeared on the "critical issues list," would be called into Flynn's office to explain why it was there—more than likely a rather unpleasant experience. Jeffrey Paul remembers when one of his permit specialists was having trouble removing his name from Flynn's list because of a stubborn member of the coast guard permit staff. "Flynn picked up the phone and called the admiral in charge of the entire U.S. Coast Guard to enlist his help and get the permit process back on track," Paul recalls.

Boston's time-honored practice of bulldozing through neighborhoods and backfilling saltwater marshes for highway projects had come to a screeching halt in 1970 with the National Environmental Policy Act. The Big Dig strictly follows the rules and regulations of this act at an enormous cost of dollars and time. As a direct result, however, Boston's environment and her citizens have been protected from permanent environmental damage and contamination. And from an overrun of rats.

The "power of the pen" has never been so evident as on the Big Dig. The lack of one permit could prevent thousands of workers and billions of dollars of construction from moving forward.

DEPARTMENT OF THE ARMY
NEW ENGLAND DIVISION, CORPS OF ENGINEERS
424 TRAPELO ROAD
WALTHAM, MASSACHUSETTS 02254-9149

3 1 JAN 1992

REPLY TO
ATTENTION OF

C# 7102
AL-1.3
MASS. DPW
I90/I93 PROJECT
FEB 5 12 10 PH '92

Regulatory Division
Operati
CENED-C
1991013

Peter Z
Central
Massach
One Sou
Boston,

Dear Mr

In
permit
undated
sheets

A
This si
above
density
lower l
The two
The cov
mil thi
of a
sedimen
inches
basical

Th
Disposa
consist
rock.

Th
permit:

a.
the cos
materia
Boston

b.
protect

U.S. Department
of Transportation

Federal Aviation
Administration

JUL 3 1 19

MHD
AUG 5 1 33 PH '92

New England Region

12 New England Executive Park
Burlington, Massachusetts 01803

J92 - 4215
C# 11698
2L-1.10

Mr. Peter M.
Massachusetts
Central Artery
One South Sta
Boston, Massa

Dear Mr. Zuk

Your request
have been rev

90-ANE-354-
90-ANE-364-
90-ANE-368-
91-ANE-024-
91-ANE-346-

The revised
#5), has resu
study is still
as possible o

If there are

Sincerely,

Barbara
Barbara L.

DEPARTMENT OF THE ARMY
NEW ENGLAND DIVISION, CORPS OF ENGINEERS
424 TRAPELO ROAD
WALTHAM, MASSACHUSETTS 02254-9149

0 8 JUL 1992

REPLY TO
ATTENTION OF

Regulato
CENED-OI
1991013

Peter Zu
Project
Massachu
One Sout
Boston,

Dear Mr.

In
modified
Departme
(Mod 12)
HARBOR
designat

Pu
transpo
East Bo
Spectac
propose
the cre
the lan

Th

BOSTON

The Commonwealth
Department of Envir
Metropolitan Boston
5 Commonwe
Woburn, Mass

Daniel S. Greenbaum
Commissioner

Atty. Joh
Brown, Ru
One Finan
Boston, M

Blossom H
177 Webst
East Bost

Dear Atty

The

Commonwealth
Executive Office
Department
Environn

William F. Weld
Governor

Daniel S. Greenbaum
Commissioner

51

CHAPTER FIVE

The Ted Williams Tunnel

HE WORLD TRADE CENTER, the St. Louis Gateway Arch, John F. Kennedy International Airport, the St. Lawrence Seaway, the Statue of Liberty, and the Ted Williams Tunnel all share the same claim to fame: Each won the American Society of Civil Engineering's highest honor, the Outstanding Civil Engineering Achievement Award—an honor that is to engineering what the Oscar is to film. The Ted Williams Tunnel was selected for this honor in 1996 because of its "contribution to the community's well-being, resource-fulness in planning, design challenges and innovative construction methods." It's a piece of work like none other.

The Ted Williams Tunnel, shortly after it opened in 1995. It is longer than the Lincoln Tunnel and almost as long as the Holland Tunnel. Its mechanical systems (ventilation fans, electronic monitoring systems) are the most advanced in the world.

The Ted Williams Tunnel reflects two passions of Bostonians—tunnel building and baseball.

It was, of course, named for Boston's most famous

baseball player, Red Sox slugger Ted Williams, at the suggestion of a baseball-loving governor, William Weld. At 8500 feet, it is longer than the Lincoln Tunnel, which burrows under the Hudson River to connect New York and New Jersey, and almost as long as the Holland Tunnel under the same river. At full capacity (95,000 vehicles per day) in 2002, it will move more traffic than either of the New York/New Jersey tunnels, carrying traffic between East Boston and South Boston and providing desperately needed access to Logan airport. To overcome two decades of political battles, environmental issues, and natural obstacles, this 1.6-mile tunnel was created with a most unusual collection of design and construction techniques. The completion of its construction was the first major Big Dig milestone after four years and close to two billion dollars in costs.

BOSTON IS BIG ON TUNNELS

In 1897 Boston beat New York City and Chicago in a race to build the country's first subway system, when the Park Street Station opened in Boston Common. At the time, most people thought putting street trolleys underground was pretty strange, but people quickly adapted to the idea and began riding the subway as part of their daily ritual. In the first year of operation, the Boston subway system carried more people than all aboveground trains combined. This was the start of Boston's love affair with tunnels.

Boston's Blue Line subway (named Blue Line because it passes under the water) was built in 1904 and connects East Boston to downtown, running under Boston Harbor. In 1934 Boston opened the Sumner Tunnel, one of the first underwater vehicle tunnels in the country.

The notorious force behind the elevated Central Artery, William Callahan, built Boston's second vehicle tunnel, the Callahan Tunnel. The tunnel's two lanes opened in 1961, and today the Sumner and Callahan tunnels run directly alongside one another. Thirty years later Boston broke ground on its third vehicle tunnel, the Ted Williams Tunnel.

and close to two billion dollars...

SCRATCHED DESIGNS, CHANGED PLANS

Strange as it sounds, the Big Dig will not reach its true final design until the day it is finished in 2005. Up until the last day of construction, engineers will review their plans and tweak the project's design. Change is constant with megaprojects like this one. The Ted Williams Tunnel is a prime example of this design flux. During six years of planning, the tunnel took on many different shapes, sizes, and directions.

One of the first changes was nothing less than the entire construction technique. At first designers considered boring the marine part of the tunnel through the earth under Boston Harbor, using a huge boring machine much like the one used to build the English Channel Tunnel. Unstable soil conditions made this impractical, though. So the bored highway tunnel gave way to a more efficient plan, using premanufactured sections called Immersed Tube Tunnel (ITT) sections.

Then there were the debates about the size of the highway, which was planned as four lanes as it tunnels under East Boston. But if four lanes were good, weren't six better? The downtown business community, among others, thought so. But Salvucci turned down this idea. Not only was it a breach of the approved design, but it would have meant a broken promise to the East Boston community—which had never agreed to a superhighway. Chalk up one victory for East Boston.

There were many other changes and debates. Toll plazas were relocated. The route of the tunnel was altered drastically, swinging in a wide curve on its approach to the airport, to avoid dynamiting near South Boston's General Ship Yard. And approaches originally designed to be open on top were roofed, or decked over. This last change had a dual advantage. If the approaches had remained open, the government would have had to compensate the property owners for air, ground, and subsurface rights. But if the approaches are decked over, the property can be returned to its original owners. This decked-over space will be developed into hotels, office towers, a convention center, and retail establishments. As a direct result, the 1000-acre seaport area will considerably increase Boston's business district, all valuable waterfront property, while the highway passes below ground, out of the way and out of sight.

LET THE DIGGING BEGIN WITH THE SUPER SCOOP!

On December 19, 1991, Governor William Weld smashed a champagne bottle against the Dutra Company's massive Super Scoop, thus beginning the construction of the Big Dig in Boston. The Super Scoop's mission: to begin digging a 50-foot-deep, 100-foot-wide, 3/4-mile-long trench below the surface of Boston Harbor. The groundbreaking kicked off a $227,000,000 contract to dredge 900,000 cubic yards of earth from the bottom of the harbor. This part of the project also included the construction of 12 enormous 325-foot-long, 7500-ton, Immersed Tube Tunnel sections. Eventually these 12 sections would be laid end-to-end in the Super Scoop's trench at the bottom of Boston Harbor.

Construction of the Ted Williams Tunnel was divided into three separate contracts. To speed up work and make the building process more manageable, the digging of the harbor trench and construction of the 12 ITT sections were placed under one contract: "the marine tunnel." The approaches to the marine tunnel, "the land tunnels," were built at opposite ends of the marine tunnel, underground in East Boston and South Boston, and were on separate contracts.

Construction on the marine tunnel began before the construction of the land tunnels. The Bethlehem Steel shipyard near Baltimore, Maryland, had won the contract to build the ITT sections and began the first construction work on the project. Shortly after work started in Baltimore, the Super Scoop

Opposite and overleaf: For two years the Super Scoop was a fixture in Boston Harbor as it dredged a trench for 12 ITT sections. Equipped with a kitchen, beds, and shower, it was "home" to its crew. The operator was guided by lasers and an onboard computer as he scooped muddy material from the floor of the harbor, 50 feet below.

started digging in Boston Harbor. This was the first physical work on the Big Dig to take place in the city. With environmental permits in hand, the very sensitive operation of removing contaminated materials from the harbor floor began.

THE POISON PIT OF DEATH

Environmentalists call it Black Mayonnaise. Others just call it muck. It's the top five feet of polluted harbor floors, made up of soft sediment filled with gunk and heavy metals—the fallout from centuries of ship traffic. The material is soaked with dark oils and grease and must be treated with care and disposed of properly. Where to put this contaminated material became one of the Big Dig's earliest and most pressing problems. The Army Corps of Engineers would not permit the material to be dumped at the Massachusetts Bay Disposal Site in the Atlantic Ocean, 20 miles off the coast of Marblehead. Desperate to find a home for the materials, project officials turned to the Massachusetts Port Authority (Massport). Massport was planning on expanding its operation into a salty mudflat called Governors Island (which is no longer an island but is now a peninsula, thanks to years of land filling).

Big Dig managers suspected, correctly, that Massport would agree to take the harbor material to help with its expansion. Of course it was not that easy. The East Boston community, near Logan airport, became alarmed after hearing false reports that the material was highly toxic, even poisonous. The Big Dig's disposal site on

Governors Island quickly became known as the Poison Pit of Death. Exaggerations coupled with legitimate concerns nearly derailed the project's first contract. Intense debates and plain yelling and shouting characterized numerous community meetings. According to one project official, the encounters were "so butt-ugly you wondered if you were going to get out of the meetings alive."

Environmentalists explained to the community that this mayonnaise-like material was no more harmful than the dirt under their city sidewalks. Finally, after tempers cooled, the neighborhood agreed to certain disposal procedures. The Poison Pit of Death was double lined with plastic sheeting and treated with lime to soak up the moisture and avoid runoff into the harbor. The "black mayonnaise" dumped into it was then treated with a special spray to prevent any part of it from becoming airborne. Then another plastic liner was placed over the top of the disposal site and clean fill was used to cap it. At the completion of the operation, the project's director of public affairs, former state senator Jack Quinlan, quipped, "The material out there on Governors Island will remain safely there until the Second Coming."

After the black mayonnaise was removed from the top of the 3/4-mile-long trench, the Super Scoop went to work on the next layer to be removed, clays and glacial tills. Barge after barge of material was towed out to the Massachusetts Bay Disposal Site, which would accept these loads.

BLASTING THROUGH ROCK

On April 12, 1992, a part of the area comprising Subaru Pier in South Boston was blown up to create a passage for the 12 ITT sections, which would begin to arrive from Baltimore in five months. The blast created a pile of rock, gravel, and mud six stories high at the edge of the harbor before it was removed. And this was just the beginning. More than two-thirds of the marine tunnel

as the Poison Pit of Death.

passed through argillite. Never before had a tunnel with ITT sections been built through so much rock. Plans for the harbor trench required the removal of 20-foot-thick sections of bedrock. In order to remove this with more explosives, the project had to jump through a hoop that made the Poison Pit of Death look like child's play.

SCARING FISH WITH $1,000,000

Blowing up a pier was one thing; but when the blasting moved directly into the harbor, the project was forced to develop a $1,000,000 "fish startle system" program to keep migrating fish away from the blasting zones. In spring, between the months of April and June, alewives and blueback herrings move through Boston Harbor from the ocean to freshwater spawning sites. Lobsters are active along the harbor floor in July, and juvenile herring make their inaugural trip through the harbor on their way out to sea between September and October. The fish traffic was on a collision course with the blasting of bedrock.

The objective of the fish-monitoring program, according to an official document, was to "detect the movement of herring in the vicinity of the blasting activities and to develop a deterrence system to move herring out of the potentially lethal portion of the blast zone." To put it in simple terms, the trick was to find the schools of fish, and move them out of the blasting area—quickly.

The drill barge carrying the boring equipment that inserted explosives deep into the rock also carried electronic systems to detect schools of fish and scare them away. The drill barge and a small boat hired to watch over the fish were equipped with electronic side-scanners to detect fish activity. Eight expensive noisemakers emitted a 200-kHz pulse to frighten the fish and force them to run from the blast area. Even with these efforts, fish would be killed, so only three blasts a day were allowed during periods of heavy fish traffic.

Ten minutes prior to a blast, the noisemakers were turned on to keep the

fish at least 500 feet from the zone, a safe distance. At this time the drill barge with its fish-scaring equipment, still making noise and looking for fish, moved out of the blast zone to prevent damage to its own drilling apparatus. The small boat stayed behind and continued to send out scary fish sounds. Two minutes before a blast, the drill barge sent a warning signal to the small boat to clear the blast area. The small boat moved to safety behind the drill barge, while both continued to frighten away schools of fish.

Boom! After the blast the small boat checked the blast area for dead fish. The environmental permits stated, "If excessive numbers of dead fish are observed following a blast, blasting operations will be suspended pending an evaluation of the effectiveness of the deterrence system." Caged fish, placed 500 feet from the blast zone, were also pulled out of the water and studied so side effects on the fish from the blast could be conducted. Boston's fish got tender loving care—or else!

Back and forth over two years of continuous dredging, the Super Scoop worked day and night with environmentalists, the United States Coast Guard, and the Federal Aviation Administration (FAA) at Logan airport watching over it. During storms and periods of thick fog, the FAA often sent the Super Scoop in to dock, with its giant crane boom lowered to protect the flight path to the runways near the harbor. Night and day the Super Scoop's aviation lights and flags were displayed as required by airport officials.

The scoop's rig was equipped with bunk beds, showers, and a kitchen for the crew. It had its own shuttle boat to transport workers off and on the barge. Computers, sonar, and monitor screens allowed the Super Scoop's operator to accurately remove 900,000 cubic yards of rock and clay from the harbor floor. The weather, fish, and oversight agencies permitting, the scoop went about its job, lowering its clamshell bucket to the bottom of the harbor, dredging a trench for the 12 massive steel ITT sections.

A FLOATING ASSEMBLY LINE

Starting in September 1992, the 325-foot, 15,800,000-pound floating steel sections began heading north from Maryland at the rate of one every month until all 12 were delivered to Boston by September 1993. As the sections of tunnel arrived from Baltimore, they were docked in Boston's Reserve Channel at the Black Falcon Terminal. When each new steel hull arrived, it was placed at the end of the dock, which was the first stop along an assembly line. Not until millions of pounds of concrete and more steel were placed inside a floating tunnel

Opposite:
Scary fish noises are made to drive schools of migrating fish away from the underwater blast area during the dredging operation of the Ted Williams Tunnel. The small lobster boat to the lower left has been converted into an electronic fish detector and noisemaker. The large drill barge (center) has just detonated an underwater explosive. The seagulls are waiting for victims to float to the top of the water.

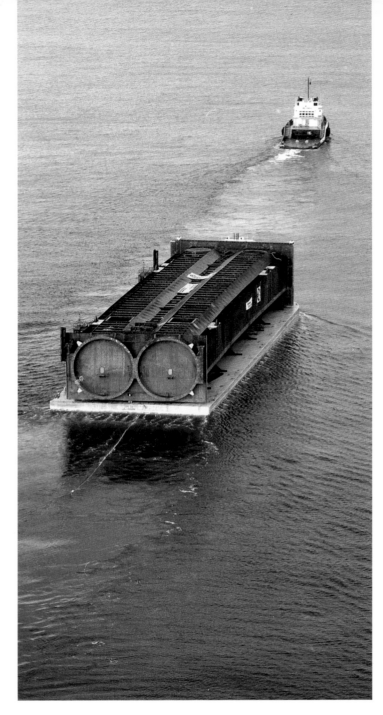

Right:
One of 12 ITT sections being shipped from the Bethlehem Steel shipyard in Baltimore to Boston on a special barge.

Opposite:
Workers on the first ITT section to arrive in Boston attach the official Big Dig logo.

section would the hull be ready for its float-out into Boston Harbor. The other ITT sections, in various stages of receiving steel and concrete, were then moved farther down the assembly line.

The ITT sections started out floating light and high in the water. As the highway tunnel started to take shape inside the hollow sections, they became heavier and heavier. After the ITTs had traveled down the assembly line, each section had gained 25,000 tons in concrete and steel and weighed a whopping 33,000 tons.

First the concrete and steel reinforcement bars went in, to give the vehicle tunnel inside of the ITT its shape and strength. Then the second phase began— the construction of the highway itself. The highway base and road deck were poured, followed by the walls and roof of what is now the Ted Williams Tunnel. To avoid capsizing the steel sections as they floated at dockside, each pour of concrete was carefully placed. If, for example, a concrete pour was made in the front right side of an ITT section on Monday, the next pour, on Tuesday, would be in the back left corner. It's a

lot like packing gear in a canoe for a trip down the river. Everything needs to be balanced so the canoe won't capsize.

FROM DOCKSIDE TO THE BOTTOM OF THE HARBOR

By the time the ITT sections reached the end of the assembly line, the steel hulls were ready to be immersed to the bottom of the harbor. At this point, the ITT sections' interior work was complete and the walls and roadbeds were clearly visible inside.

A laybarge—a catamaran-like apparatus—was maneuvered into position and straddled the floating ITT section with its pontoons. When the barge's two pontoons, affixed to two large steel girders, were atop the ITT section, it attached itself to the floating hulk with ropes and cables. The laybarge held the ITTs above water as the last concrete was poured into compartments on the sides, in the center, and on the top of the floating section.

Five ITT sections docked at a pier in Boston Harbor and floating on their own. The ITT section to the far right is heavy and sitting low in the water. The most recent arrivals are to the far left and have yet to be built out with an interior concrete tunnel.

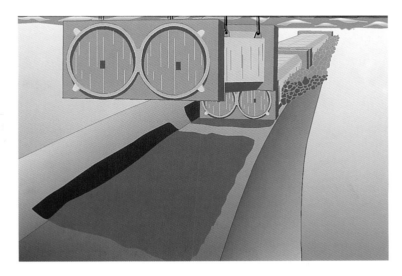

This created negative buoyancy, so that the giant sections would sink on their own. Up until this point, the ITT sections enjoyed the physics of displacement and resisted immersion. Now they were held up by the laybarge with almost 12,000 pounds of negative weight.

The first tube section was ready to be lowered in February 1993, five months after its arrival into Boston Harbor. Just before the inaugural launching and before an opening in the section's roof was sealed forever, a cement truck was brought out to the dock and taken apart. A crane then picked up the parts and lowered them through a small hole in the roof of the ITT section. After all the truck's parts were inside, the truck was reassembled and put to use for two years at the bottom of the harbor, shoring up gaps between the 12 sections.

After that first tunnel section was placed into the trench, the ITT sections moved down the assembly line and were lowered into the harbor's

Three computer images of an ITT section being lowered into place and connected to another ITT section. Twelve ITT sections were placed in a 3/4-mile-long trench from East Boston to South Boston, dug by the Super Scoop. Granite boulders are placed over the top for protection.

bottom at the same rate at which they had arrived from Baltimore—one a month. By fall 1994 all 12 were in place. Now for the final connection.

DIVERS, LASERS, SATELLITES, AND NOT AN INCH TO SPARE
Underwater construction divers working on the Ted Williams Tunnel at the bottom of Boston Harbor had less than 12 inches of visibility under good conditions. They often closed their eyes while working, preferring to use their sense of touch. Frigid ocean temperatures required that hot water be pumped through the divers' suits so they could remain at the bottom of the harbor long enough to complete a task. All divers were equipped with lights on the exterior of their diving helmets and radio transmitters on the inside. The radio kept the divers in constant contact with the dive barge 100 feet above.

The United States Coast Guard gave Big Dig officials 24 hours to place each ITT section in the Super Scoop's trench. The laybarge, with an ITT section attached, was moved out of the Black Falcon Terminal, through the harbor's commercial channel, and into place over the 50-foot-deep trench. A laser beam was shot across the harbor between South Boston and East Boston, creating a guide for the laybarge as it maneuvered into place. The laser light helped the technicians find the centerline of the deepwater ditch 100 feet below. Global Positioning Systems (GPS) were deployed to search and find the exact location of the ITT relative to its destination at the bottom of the harbor. Each of the 12 ITT sections was placed within a fraction of an inch of forgiveness on top of a bed of gravel.

At the bottom of the trench, four large couplers, similar to those used to connect boxcars on a freight train, reached out from the ITT section. Four hydraulic jacks extended the couplers from each corner of the section and made connections with the couplers of an ITT section already in place. The hydraulic mechanisms pulled and locked the couplers together until

Previous page:
An ironworker hangs onto an ITT section about to be sent to the bottom of Boston Harbor.

Opposite:
A welder pauses while working atop one of the 12 ITT sections in South Boston. Whether summer or winter, welders often cover themselves with suede that acts as a fire retardant from red-hot sparks.

the divers' suits so they could remain at the bottom of the harbor...

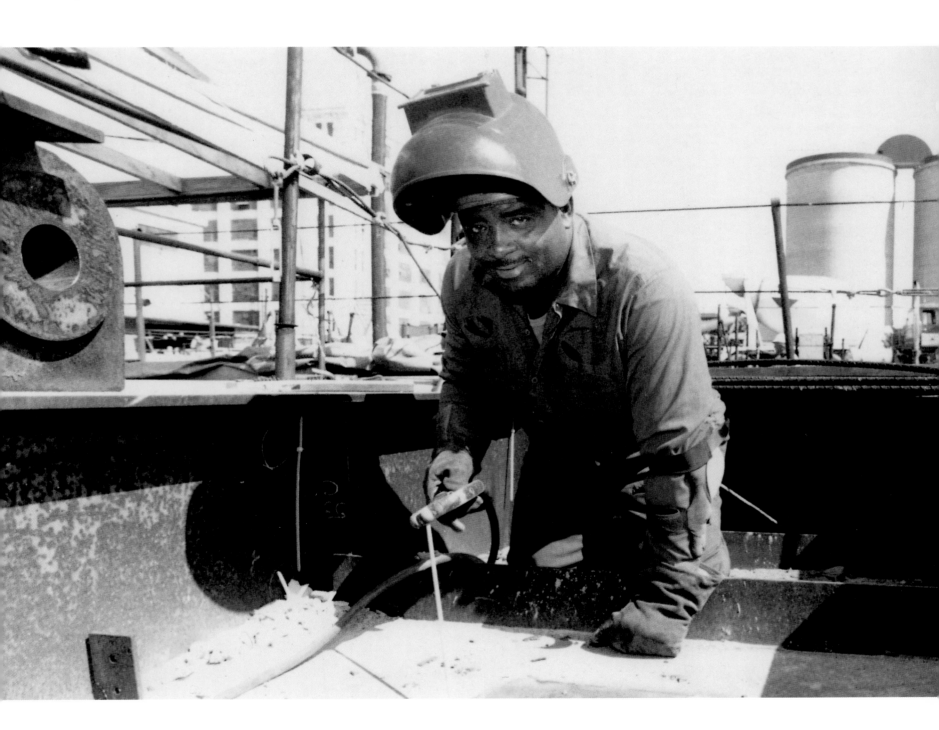

A worker stands inside one of the two barrels of an ITT section as it floats in the water in South Boston. Directly above his head is a glory hole used for lowering building materials and equipment. Before the tunnel section was immersed to the bottom of the harbor, the glory holes were sealed shut and the interior of the ITT section was kept bone dry.

massive rubber gaskets extending from each end of the connecting ITT sections were compressed. Divers then checked the connections of the ITT sections to make sure there was a secure, airtight seal.

Once the divers gave the okay, the final step of connecting the 12 ITT sections began. Each section was sealed at both ends to be airtight. The gap between each ITT section was all that had been exposed to the underwater elements. Every time a connection was made, water, sand, and rock were trapped between the enormous tunnel sections. This material was sucked out with pumps, pipes, and valves until thousands of gallons of water rushing out of this area created a hydrostatic seal—an air lock that permanently connected the two separate ITT sections.

Upon linking the ITT sections together and after the water and debris were removed, air-lock hatches similar to those on submarines opened between each ITT section. Once this important link was established, tunnel workers ventured inside to weld the sections together to ensure the joints were watertight. Then the ironworkers turned their torches on the three-inch-steel bulkheads at the ends of each section. These bulkheads were cut into five-foot pieces over a period of two to three days and taken out of the tunnel. The gaps between the ITT sections were filled with concrete and steel to make the tunnel seamless. It took two years of work before the last two bulkheads were removed from the extreme ends of the tunnel. Not until the summer of 1995 were workers able to walk through the 12 ITT sections, unobstructed, from South Boston to East Boston.

A carpenter works inside a floating ITT section in South Boston. The long ladder in the background is the only way in or out.

The last two bulkheads to be removed from the 12 ITT sections were in East and South Boston. A worker on a "man-lift" inside the land portion of the Ted Williams Tunnel cuts away the bulkhead in South Boston, finally connecting the 12 ITT sections with their approaches. A bulkhead door, similar to a submarine's, can be seen directly below the welder.

ONE BY SEA, TWO BY LAND

Construction of the two separate land tunnels leading to the marine tunnel began in the summer of 1992. Work started on the South Boston approach in July and East Boston's in August.

Logan International Airport is the sixteenth busiest airport in the country and one of the smallest, sitting on about 2500 acres of man-made material in the middle of Boston Harbor. The Massachusetts Port Authority, which owns and operates the airport, is very strict about access to its property. The biggest concerns for Massport are the safety of passengers and jets, security at the airport, and service to its clients, the airlines. The least of its concerns was the daily progress of a tunnel under the harbor. Massport was not happy that the Big Dig was ripping apart its property.

The construction of the East Boston approach, a six-lane tunnel, was complex, requiring 60 tons of reinforcement bars and 450,000 cubic yards of concrete to make 100-foot-deep cuts. The approach plowed through weak soil that was 100 percent man-made, the result of landfill operations in East Boston between 1915 and 1975. Because of this, a Japanese method of deep-soil mixing, never used in the eastern part of the United States, was employed to construct temporary support walls to secure the unstable soil. Additionally, the roof of that deep highway was poured with four extra feet of concrete to support jet aircraft on a taxiway above.

Workers in the early stages of building two lanes of roadway inside a floating ITT section docked in Boston Harbor.
To the far left, two more lanes of road are under construction.

Opposite:
A giant auger frees itself of dirt from a deep excavation in East Boston. Used to drill shafts on the Big Dig, these augers are as large as ten feet in diameter and travel 150 feet into the ground.

An old Eastern Airlines terminal obstructed the path of the tunnel, which would be nearly 100 feet below the airport tarmac. A hangar was torn down and a jet-blast wall, which had protected people from air shooting out of jet engines, was relocated. Because of space limitations, larger jets were restricted from using the terminal during three years of construction. Dump truck convoys constantly rumbled along three active runways. Finally, the Big Dig's contaminated dirt pile, the Poison Pit of Death, almost required the relocation of the airport's primary radar system.

These were not happy times for Massport. It demanded that the Big Dig's disruptive work be performed under its own strict guidelines. Cranes, which were already forced to fly flags and display flashing aviation strobe lights, were told to "boom-down" at the hint of bad weather. Since this required that the boom be lowered to the ground, all work at the airport came to a grinding halt. A feud of sorts developed between Massport and the Big Dig.

Massport demanded that trucks seeking access to the runway area be marked with large SECURE signs. Airport officials stipulated that security check-points and guard shacks for the Big Dig trucks and workers be built to precise specifications. All trucks were checked with mirrors and bomb-sniffing dogs before each day of work. Dump trucks were forced to travel in a convoy to a temporary dumping site near one of the runways and were required to have an airport security vehicle escorting them at all times. Tensions escalated and interfered with progress.

The disagreements could easily have destroyed the critical path of this part of the dig and delayed the project's first major milestone—the opening of the Ted Williams Tunnel. Fortunately, Massport officials and Central Artery/Tunnel Project managers sorted things out and were able to work through their respective needs.

THE LARGEST COFFERDAM IN NORTH AMERICA

In South Boston the same type of work was being performed. An approach from land was being built to connect with the 12 ITT sections under the harbor. The soil here was like East Boston's, weak and man-made, and groundwater fluctuations complicated construction. Surges from the harbor's tides force upward anything placed in the earth nearby. This meant the highway needed to be anchored so it wouldn't "float."

To counter this, project engineers designed the approach road with a 17-foot-thick surface called a gravity slab, poured over waterproof membranes. This thick foundation stretched for a half mile and was tied to the bedrock 100 feet below with enormous cables. But that wasn't all that was needed to stop Father Neptune from overwhelming the tunnel.

The land tunnel with its cross-lot struts in South Boston meets with the circular cofferdam. The dam holds back the ocean, allowing the ITT sections to link up with the cut-and-cover tunnel. Inside the cofferdam is the beginning of a ventilation building. Logan airport is in the distance across Boston Harbor.

The largest circular cofferdam ever built in North America was constructed to connect the land tunnel in South Boston to the 12 ITT underwater sections in Boston Harbor. The dam, 85 feet deep and 250 feet in diameter, restrains the ocean during high and low tides with its 13-foot-thick walls, which are buried into bedrock.

Inside the circular dam two structures were built. First, a 250-foot-long section of highway tunnel was constructed to connect the land tunnel with the marine tunnel. After the tunnel section was finished, a ventilation building was built on top of it. In the summer of 1995, construction vehicles could finally travel from South Boston through the cofferdam and into the 12 ITT sections to East Boston and the airport. The tunnel was nearly finished.

OPEN FOR BUSINESS

Close to each end of the tunnel's openings, thousands of gallons of salty groundwater was leaking into the approaches, causing newly placed tiles to fall off the walls. Ugly yellow streaks began appearing on the new tunnel walls. Ironically, it was the land-based sections, deep in the groundwater under East and South Boston, that were leaking so badly. The ITT sections 100 feet below the surface of Boston Harbor were watertight and free of leaks.

After a great deal of debate and finger-pointing, work crews began injecting thousands of gallons of grout into the ground along the tunnel walls. This kept the water from seeping into the tunnel's interior. Finally the water was controlled, and today the 1,400,000 eight-inch hand-placed tiles are solidly in place and free of nasty yellow streaks.

On the snowy eve of the tunnel's opening, crews were still working on the finishing touches—checking the 34 ventilation fans, fire systems, lighting, and water-pumping systems. The white lines were just being applied to the road surface.

Ted Williams, known as "The Kid," during his legendary career, waves to the camera as Governor William Weld (left), and his son, John Henry Williams, stand on either side of him immediately after the tunnel's opening day ceremonies on December 15, 1995.

Almost four years to the day after the Super Scoop arrived in Boston Harbor to begin dredging in the harbor, the tunnel was about to be opened.

At 10:00 a.m. on a cold and cloudy Friday, December 15, 1995, the United States Marine Corps Band led a vintage convertible out of the new tunnel exit in South Boston carrying the city's most famous baseball player ever, Ted Williams. Governor Bill Weld was behind the wheel, and along for the historic ride were former governors Ed King and Michael Dukakis. At last they were able to make the trip from East Boston to South Boston in a tunnel that had taken over 30 years of political battles and engineering genius to complete. As Ted Williams said of the tunnel, "You'll be talkin' about it for a hundred years."

CHAPTER SIX

Spectacle Island

N TINY SPECTACLE ISLAND, Big Dig archaeologists uncovered garbage from ancient Indian campsites dating back 1500 years. On the island, along with debris from ancient Indian activity, were 97 acres of leftovers from 350 years of pleasant and not-so-pleasant occupation. The island was covered with remnants of a slaughter-house, a grease-extraction plant, car parts, and millions of tons of garbage reaching 100 feet high and extending 550 feet off its shoreline.

The ugly duckling of the Harbor Islands, Spectacle Island is seen here in the early stages of its comeback. A garbage incinerator's red brick stack is standing to the far right and in the foreground is the island's north drumlin.

The Big Dig, with $165,000,000 from its deep pockets, has rescued this environmental disaster zone from horrific ecological abuse. Nearly 4,000,000 cubic yards of material, 25 percent of the dirt to be unearthed by the Big Dig, has been placed on Spectacle Island. A 60-foot-thick cap of clay and gravel fill from the Ted Williams Tunnel and other downtown construction sites creates an impermeable seal atop the 81-year-old city dump. The island's marriage with the Big Dig reads like a fairy tale—an ugly duckling becomes a beautiful swan. This huge garbage dump, transformed into a recreational facility, becomes part of the new Boston Harbor Islands National Park Area in 2003. As Brian MacDonald from the New England Aquarium stated on a recent trip to the island, it is the "greatest story you can tell about recovery."

SMALLPOX AND BROTHELS, DEAD HORSES AND METHANE FIRES

Information must be salvaged from prehistoric archaeological sites that are destroyed by projects like the Big Dig. The National Historic Preservation Act of 1966 mandates that if a site cannot be avoided or preserved, historical material from that location must be gathered, studied, and recorded before construction begins. So in the summer of 1992, before tons of earth from the Big Dig buried an ancient Indian hunting ground forever, a small group of archaeologists arrived laden with gear to conduct their own dig.

After forcing mountains of earth through one-quarter-inch holes in metal screens, archaeologists uncovered a cache of artifacts documenting an active life on the island thousands of years ago. They brought the haul back to their laboratory for a closer look and radiocarbon dating. The archaeologists removed a total of 225 artifacts such as pots, tools, and weapons, along with 6555 bones and 1560 pounds of shells. What they learned from these splinters of prehistoric time is considerable.

Eons ago a mile-high glacier, crushing and grinding down sharp peaks and ridges, moved through the Boston area and created small rounded hills, or drumlins. Beacon Hill, Bunker Hill, and the Boston Harbor Islands are famous examples of these glacial drumlins. The glaciers melted 16,000 years ago and eventually Boston Harbor was formed. In time it became a popular fishing area for the Algonquian Indians. The language of the Algonquians is still with us in names like Nantucket, Connecticut, Narragansett, and Merrimack.

The most ancient piece the Big Dig team found on Spectacle is a 7000-year-old arrowhead, consistent with other archaeological findings around the harbor islands but random among the rest of the artifacts dug up on Spectacle. Except for this one arrowhead, the entire load of material discovered dates back 1040 to 1415 years.

The Indians who left their mark on Spectacle Island were most likely from the Neponset Valley, 27 miles away up the Neponset River. They came to the island in dugout canoes during the spring and fall to hunt and fish but didn't stay for long periods of time. The campsites' debris indicates that they ate white-tailed deer, raccoon, and even dog. They also fished and dug for clams, ate much more cod than flounder, preferred soft-shell clams over blue mussels and quahogs, and may have eaten turtle since turtle shell fragments were found. Archaeologists concluded that the Indians cooked hickory nuts that they brought with them—there are no traces of hickory trees on the harbor islands—and smoked herbs in artfully decorated pipes.

The first colonists from the New World chose to settle in the port of Boston in the early 1600s because of its safe, protected waters. The 30 Boston Harbor islands created a natural shield against violent ocean storms and enemy attacks—a major concern for a settlement in the New World. In 1634, colonists noticed that one of the islands had two drumlins connected by a sand bar and thought that it resembled a pair of eyeglasses or spectacles floating in the water. Thus the name Spectacle. In 1684 Samuel Bill bought the island from the son of Wampatuck, the chief of the fast-diminishing Massachusetts Indians.

By 1707 a small house stood on the island. Spectacle had been cleared for timber and a pasture before a quarantine hospital was erected in 1717.

Over $3,000,000 is spent on the completion of the Big Dig every day during the peak construction years, 1998-2001.

The place was literally overflowing with refuse when a bulldozer

This facility turned into one of the first immigration-processing centers of the New World, a predecessor to Ellis Island. Ships arriving from Ireland into Boston as early as 1729 were forced first to pull into Spectacle Island and unload passengers who were plagued with smallpox, yellow fever, and other diseases, before sailing into the city.

Prior to the Civil War, Spectacle was a favorite destination for Bostonians taking Sunday boat trips and summer picnics. It became so popular that two seasonal hotels were opened in 1847 on top of the island's north drumlin, until the holidayers got carried away and the inns turned into gambling halls and places of ill repute. Police raided the island in 1857 and shut down a brothel, which in turn led to the closing of both hotels.

Immediately after the police raid ended the frolicking on the island, a Mr. Nahum Ward bought it for $15,000 and set up a horse-rendering plant. Mr. Ward explained, "Every day the steam-tug and barges pertaining to the company go down from their wharf on Federal Street, laden with dead horses and refuse from slaughter houses. If they were allowed to remain in the city for three days in the summer they would cause disease." By 1882 his island supported the plant, 30 men and 13 of their families, and a small schoolhouse. But even this didn't last. The slaughterhouse closed in 1910 when the electric trolley replaced horses as the main source of urban transit. The island quickly sank to its lowest point when the city began to dump garbage on it two years later.

In 1921, a grease-extraction plant began operation on Spectacle. Garbage was barged out from the city, cooked, compressed, and then the grease removed and sold as an ingredient in the manufacturing of soap. The plant operators disposed of the by-product, burned, flattened garbage, by dumping it on the island and covering it with rubbish. After the plant closed in the 1930s, the city continued to dump waste on the island until an additional

Opposite:
The New York Times called Spectacle Island "…a 100-acre hell-hole three miles from Boston." Before the Big Dig capped the island, 22 million gallons of toxic run-off oozed its way into Boston Harbor each year.

disappeared into a hole in the rubbish.

36 acres had been added to the island's girth and 100 feet to its height. The place was literally overflowing with refuse when a bulldozer, working at the dump in 1959, disappeared into a hole in the rubbish. The city discontinued its operations, and the island dump was abandoned and closed to the public. In the 1960s, fires from the spontaneous combustion of methane gases created by the tons of garbage broke out and burned for years.

Carey Goldberg of *The New York Times* wrote of Spectacle Island, "Think of the exact opposite of St. Tropez, or the anti–St. Lucia. Imagine a hideous, stinking, toxin-leaking dump right in the middle of Boston Harbor, which itself used to be compared to a toilet, what with sewage and all. A 100-acre hellhole just three miles out from Boston that no one was allowed to land on, and who in the world would want to anyway?" This colorfully describes the island's status until 1992, when dirt from the Ted Williams Tunnel began to arrive. Spectacle's long awaited recovery was about to begin.

PHOENIX RISING FROM THE ASHES

When World War II broke out, the Department of Defense took control of strategic islands in Boston's harbor to protect the navy's shipyard in Charlestown. To prevent an attack from Germany, the entrance of the harbor was heavily mined and sealed off with an underwater torpedo net. Anti-aircraft guns and radar systems installed on the islands protected the sky. In 1970, safe from a German invasion and willing to help Massachusetts create a harbor-island state park system, the Defense Department sold its interest in the islands to the commonwealth. However, without money for a costly rehabilitation, Spectacle Island remained unwanted, untouched, and closed off to the public.

Governor Michael Dukakis in 1986 proposed that the Central Artery/Tunnel Project create a win-win-win situation among the city of Boston, the Commonwealth of Massachusetts, and the U.S. government by forming a

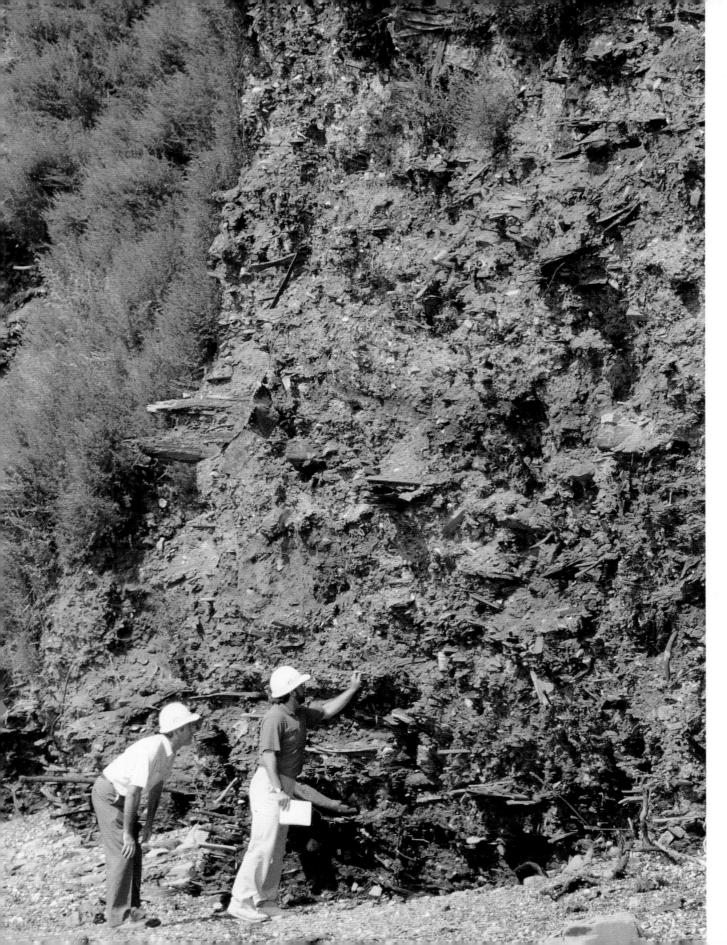

A mountain of garbage created 36 acres of additional land that was capped by the Big Dig. In the 1960s, spontaneous combustion ignited fires on the island and the Boston Fire Department decided to let them burn themselves out. The fires burned for years.

partnership to improve the island. He suggested capping the noxious garbage heap with the entire load of dirt from the Big Dig—12,000,000 cubic yards as projected at that time. This benefited the three parties involved by reducing the cost of expensive material disposal for the federal government and Massachusetts, and eliminated an environmental hazard for the city. This proposal crashed and burned; but the spirit of the city, state, and federal partnership concerning Spectacle survived.

Dukakis's plan would have doubled the size of the island from 97 to 200 acres, extending it dangerously close to Presidents Road, the commercial shipping channel into and out of Boston Harbor. The U.S. Army Corps of Engineers was concerned that a distended Spectacle would create a navigational hazard. Also, the proposal did not include the costs of building a park after the project's dirt was dumped on the island, which rallied local environmental groups, such as Save the Harbor and The Friends of Boston Harbor Islands, to file a lawsuit and kill the plan.

Four years later, final plans for the Ted Williams Tunnel were in full gear and project officials were in a jam. Desperate for a place to dispose of the millions of tons of earth scheduled to come out of the ground for the tunnel's construction, they struck a deal with environmentalists. The project agreed to pay for the essentials of an island park and to limit the dumping to 2,700,000 cubic yards of material instead of the original 12,000,000 cubic yards. This cleared the way for construction on the South and East Boston approaches of the tunnel.

MOUNT SPECTACLE

The first barges of dirt from the Big Dig arrived on Spectacle Island in the summer of 1992. Contractors had first delivered the earth to the Subaru Pier in South Boston for environmental testing. At the pier the Big Dig's earth was sorted, checked, and examined for contaminants. Once environmental regulators and project officials approved the material, they sent it to Spectacle Island. There cranes, conveyors, road graders, massive dump trucks, compactors, bulldozers, and rollers flattened and spread the dirt from South and East Boston, reshaping the harbor's ugly duckling.

While the digging was under way for the Ted Williams Tunnel, dirt from construction in downtown Boston arrived on the island. Starting in the summer of 1992, tons of backfill resulting from miles of underground utility relocation arrived. Moving gas, water, electric, phone, and thermal lines from the path of the highway tunnel resulted in an enormous number of dirt shipments to Spectacle. Barges arrived day and night continuously, seven days a week for years.

In the spring of 1995, excavated material from the first downtown construction work was making its way to the harbor island. Dirt sent from the subterranean highway from the construction of four northbound lanes under Atlantic Avenue near South Station was followed by dirt from between State Street and North Street. Within the year, the island had reached the 2,700,000 cubic-yard limit that environmental regulators had set—and to everyone's surprise, there was room for 1,000,000 cubic yards more. Material from the Fort Point Channel and South Boston construction was diverted to the island as a result, and the project saved $20,000,000 in shipping costs.

The dredged material arriving to the island was so wet that a construction manager nearly drowned in mud. Sinking up to his chest in the dark, dank material, he was trapped until discovered and dug out by fellow workers. To deal with the wet material, workers resorted to extreme measures to dry, crush, mash, and compress the millions of cubic yards of earth. A jet engine attached to a tractor blasted hot air into wet soil arriving from the Fort Point Channel and other water-saturated areas of the city.

About two-thirds of the Big Dig's dirt will be trucked away— that's 541,000 truckloads. If all those trucks were lined up end- to-end, they'd span 4,612 miles. That's all the way to Brazil, as the crow flies.

A clamshell bucket unloads dirt from a barge and shovels it onto a giant conveyor belt. 3,900,000 cubic yards of earth has been placed on the island from the Ted Williams Tunnel, Fort Point Channel, and downtown Boston.

Similar in concept to equipment used at horse racetracks after a rain shower, the device produced marginal results but reflected the creativity at work to accommodate the fill. More effective was lime mixed into the barged dirt to dehydrate the wet material. The technique was so efficient at drying the soil, Mike Virta—a Spectacle Island engineer—said, "What looked like soupy

The conveyor belt dumps earth arriving from the material processing operation on Subaru Pier in Boston. Reaching high into the air, the conveyor belt allows enormous piles of earth to form under it so large dumptrucks can pick up the earth and distribute it over the entire island.

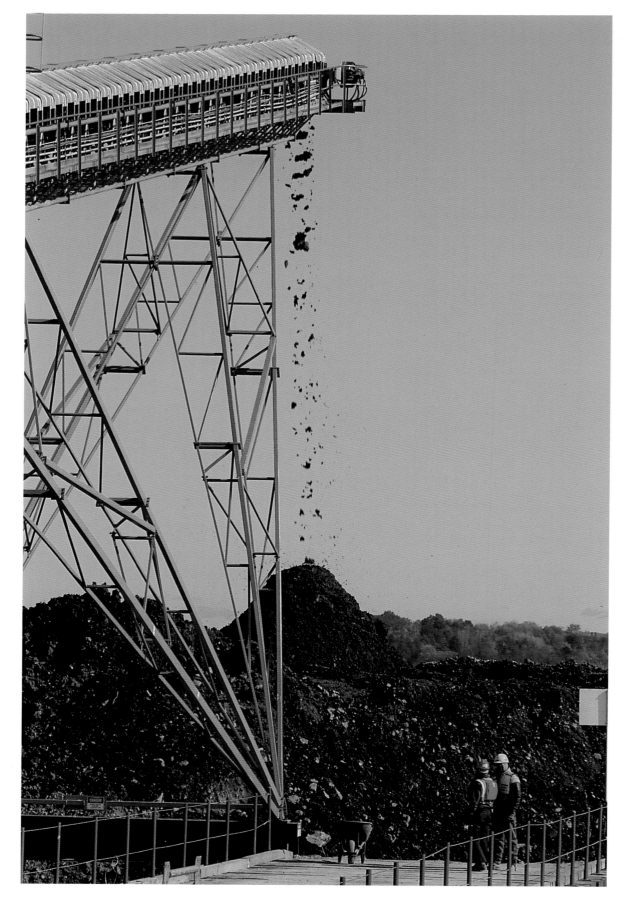

elephant droppings on arrival was effectively compressed as dry dirt the next day."

The original design specifications and height limits were not altered, even though more earth than planned was placed over the former garbage dump.

A total of 3,900,000 cubic yards was placed on the island, doubling the height of the north and south drumlins. Before long, South Boston residents were calling it Mount Spectacle.

CREATING A SPECTACLE

Three decades after the island dump closed, a stream of steaming leachate could be seen oozing out of the bowels of Spectacle Island into Boston Harbor. Containing the river of toxins—22,000,000 gallons a year of noxious metallic liquid running off the island into the harbor—was a top priority.

Brown and Rowe, a local landscape architectural firm, assisted in the environmental design for the new Spectacle Island, adding shapes, contours, and an enormous drainage system. The firm laid out a strategy for capping the rubbish heap, capturing hazardous runoff, and planting 28,000 trees, shrubs, and plants to prevent erosion and attract wildlife.

Sewing up the island and keeping it together is the job of the largest man-made structure on the island. A 35-foot-thick granite seawall protects the island from harsh winds and storm waves that prevail from the north and east. The containment dike is a mile and a half long and wraps around two thirds of the island. It holds Spectacle together and provides the foundation for a long section of walking trail. The excellent views of the other harbor islands from along this stretch of the promenade make it one of the island's most inviting features.

Before millions of tons of earth from the Big Dig was spread about the island, a drainage system was built to capture contaminated runoff from the garbage dump. Today, leachate from old refuse can funnel into the center of

Overleaf:
One truckload closer to the 6,700,000 tons of earth placed on the island. The dirt was piled six stories high to cap the nearly 100-year-old dump.

More than 3000 trees and 25,000 shrubs will eventually be planted at Spectacle Island, transforming an embarrassing eyesore into a pristine park.

the island, where a system of rocks and erosion-control blankets create an impermeable surface. If contaminated water reaches this layer, it drains through galleries to a recirculation wet well on the east side of the island. The runoff collected into the well is pumped to the top of the drumlins, and the process begins anew, taking as long as a year to cycle back through the system. This is probably the largest septic system in New England, with a capacity of 1,000,000 gallons. Environmentalists expected 2,000,000 gallons of noxious liquid to cycle through the system annually. However, after more than a year of examination, the wet wells are free of contaminated water, verifying that the island currently has an astonishing 100 percent captivity rate.

In the landscaping phase of the remaking of Spectacle, the natural shapes of a glacial drumlin island were recreated with glacial till and clay. Brown and Rowe then worked with a soil scientist to develop a five-foot-thick layer of man-made loam. The manufacture of this topsoil was ingenious. Glacial till, sand, and organic matter were mixed together on the island to create the man-made soil. Almost like alchemists of old, the landscapers made a huge quantity of loam without stripping acres of local property of the critical material. Today, many of the planned trees and shrubs have already been planted into the five-foot-thick surface, preventing millions of dollars of loam from washing away. An impressive array of plant species gives off a show of colors throughout the seasons.

Spectacle Island is already attracting the types of wild animals and creatures that fled the island during its decades of abuse. Swimming from one island to another, deer, coyotes, raccoons, fox, and muskrats have added Spectacle back to their list of food sources. From the air, ducks, birds, and even a snowy owl have visited the island. Lobsters are now found in considerable numbers near the seawall.

TAKE A HIKE

Spectacle is Boston's fourth largest harbor island and the most accessible from downtown. It stands as the gateway to the other islands, the first one a boater comes upon when steaming out of the harbor. The island's marina has an L-shaped pier, constructed to accommodate boat slips and docking facilities for large ferries. The pier design allows for a future restaurant, sailing center, and other light development. More than a half mile of sandy beaches, in two locations on the island's weather-sheltered side, could entice . . . yes, swimmers into the cold water. The harbor that for decades was infamous for its filth has made a comeback that rivals Spectacle's.

"You couldn't see the bottom in three feet of water back in '79. Now you can see eighteen feet and the harbor is filled with fish, even porpoises and seals," explained Kathy Abbot, a former harbor island park ranger, in a recent *Boston Globe* article. The water has improved dramatically, thanks to an enormous Boston Harbor cleanup project that was ordered by a federal judge. A sophisticated water filtration system on nearby Deer Island now treats and safely disposes of raw sewage that once went directly into the harbor.

As *The Boston Globe* said in May 1997, "Spectacle Island may be transformed from a dirty little secret into one of the city's more desirable recreational destinations." Sailing, kayaking, swimming, windsurfing, hiking, summer concerts, and even cross-country skiing are in the mix. Five miles of walking paths winding around the island offer a series of beautiful views, and bring hikers to the top of the island's two highest points. The south drumlin now tops out at 120 feet, and the north drumlin offers a superb 360-degree view at 150 feet above sea level, the highest point among the islands. Engineer Mike Virta says, from the top of the north drumlin, "The uses are almost endless. Every time I bring someone up here, their reaction is 'Wow.'"

THE COMEBACK ISLAND

The entire island will become a museum of sorts, a learning center displaying remnants of the gruesome acts man committed against his most precious resources and living examples of his ability to orchestrate a spectacular comeback. Industrial waste, factory machine parts, navigational systems, and 20 barrels of random debris were collected from the island before its cleanup. These samples of neglect memorialize environmental atrocities from the past in a permanent exhibit in the future visitor center.

The island itself will demonstrate a successful sustainable environment. Self-composting toilets, solar panels for electricity, battery-powered vehicles—all exemplify the conservation of natural resources. A classroom facility in the visitor center will provide a venue for discussions on protecting, not destroying, natural resources.

Spectacle Island stands as an example and a paradox: an amazing environmental comeback showcasing great and terrible deeds—made possible by the efforts of many groups and individuals, but most notably by the Big Dig.

Brown and Rowe, a landscape architectural firm, assisted in Spectacle Island's new look by adding shapes and contours. The dock, shown here in the shape of an asymmetrical cross, will actually be L-shaped.

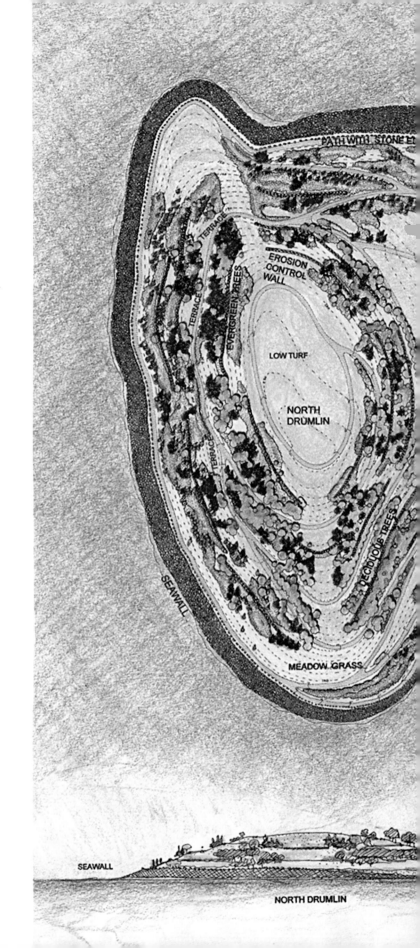

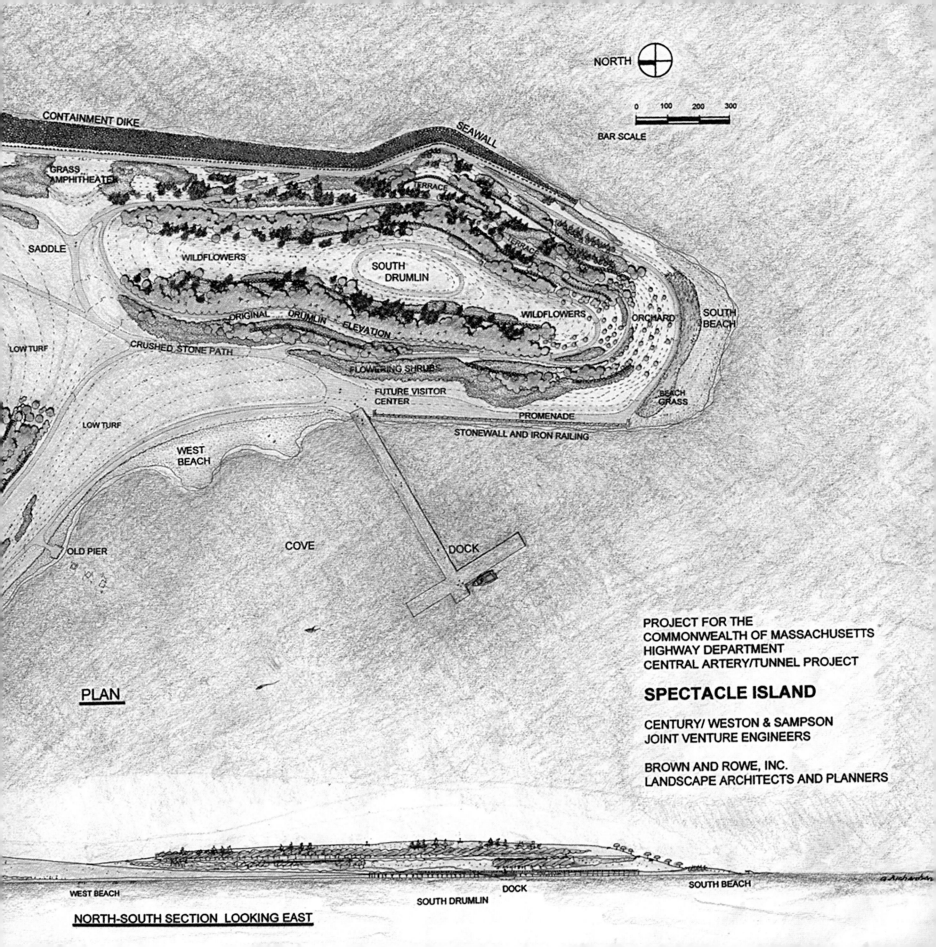

NORTH

0 100 200 300
BAR SCALE

CONTAINMENT DIKE

SEAWALL

GRASS AMPHITHEATER

TERRACE

SADDLE

WILDFLOWERS

SOUTH DRUMLIN

TERRACE

LOW TURF

ORIGINAL DRUMLIN ELEVATION

WILDFLOWERS

ORCHARD

SOUTH BEACH

CRUSHED STONE PATH

FLOWERING SHRUBS

BEACH GRASS

FUTURE VISITOR CENTER

LOW TURF

PROMENADE

WEST BEACH

STONEWALL AND IRON RAILING

COVE

DOCK

OLD PIER

PLAN

PROJECT FOR THE
COMMONWEALTH OF MASSACHUSETTS
HIGHWAY DEPARTMENT
CENTRAL ARTERY/TUNNEL PROJECT

SPECTACLE ISLAND

CENTURY/ WESTON & SAMPSON
JOINT VENTURE ENGINEERS

BROWN AND ROWE, INC.
LANDSCAPE ARCHITECTS AND PLANNERS

WEST BEACH

SOUTH DRUMLIN

DOCK

SOUTH BEACH

NORTH-SOUTH SECTION LOOKING EAST

CHAPTER SEVEN

Slurry Walls and the Maze Below

ELOW BOSTON'S BUSTLING downtown streets is an underground obstacle course of utility lines, subway lines, and interesting obstructions like old buried wharves and building foundations. Most of this, it seemed, was directly in the path of the new Central Artery Tunnel. Above and next to the excavation site are some of the most historic buildings in the country. If that isn't enough, the tunnel is to be built beneath the existing expressway bridge, while cars and trucks drive on top. Overall, this segment of the project required the application of some of the most complex, sophisticated construction techniques ever attempted. All of it carried out in tiny work zones, with cranes and dump trucks wedged into the heart of the busy city.

Two Big Dig workers wait for a 14-ton "slurry bucket" attached to a crane to reach the surface after taking a 100-foot deep trip into the ground. The massive clamshell bucket excavates one section of slurry wall at a time. The Big Dig has the largest concentrated use of slurry-wall construction in the world. Over 26,000 linear feet of slurry walls have been built on the project.

The key to accomplishing all of this is something called slurry walls. In a typical tunnel, a large hole is excavated, and then concrete is poured from the bottom up to form the walls. This plan wouldn't work in downtown Boston. The hole would need to be too wide, and there would be too much disruption to the existing buildings. Slurry walls allow for construction of the tunnel wall first—before excavating the hole. This is done by digging a narrow slot from the surface, filling the slot with a liquid clay called slurry, and then filling the trench with reinforcement and concrete. The concrete pushes the liquid slurry up and out, then hardens into a solid wall. The finished wall can then be used to support the sides of the tunnel, which is dug out between these walls. This method also allows for a narrower tunnel, useful in Boston's crowded underground, because the slurry walls can be used both for support of the excavation and as the final tunnel walls.

But before the slurry walls could be constructed, there was one little problem—the maze of underground utility lines running every which way below the city. Some of these old utilities are very old. They were not exactly around during the ride of Paul Revere, but some were in place shortly thereafter.

ANYTHING BUT SIMPLE

The problem of building a tunnel around an old utility line is that the line can crumble or collapse when the construction begins. And of course the utility line must remain in service, or downtown Boston would fall into chaos, without electricity, water, or flushing toilets. Relocating utilities is a complicated exercise in scheduling and construction finesse. Tunnel constructors needed to win this relocation game at all costs—or risk losing public support for the project.

The basic Central Artery/Tunnel utility relocation scheme is shown on the facing page. "Before" is a chaotic state, with utility lines running in all directions.

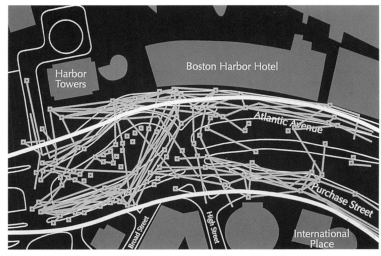

Schematics of the "before" and "after" utilities in the path of the Central Artery tunnel. "Before" (above) has a disorganized alignment of old utilities.

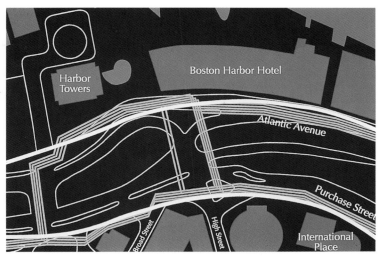

In the "after" picture, the utilities have been neatly realigned into corridors and rebuilt in preparation for the tunnel excavation and construction.

"After" shows the utility lines relocated into two basic types of corridors: running parallel on the sides of the tunnel and running a lateral slot across the tunnel. The relocated utilities are all strengthened and improved, no small job since some date from the eighteenth and nineteenth centuries. It seems like a simple exercise, at least on paper. But to schedule and arrange this work was not so simple.

To make matters more difficult, utility lines have "personalities." Each utility has its own particular quirk that must be considered when designing and constructing the rerouting. For example:

Electric lines have electricity. When planning the rerouting, it was important to consider how close these lines would be to cutting-and-scraping machinery and operations. Cutting through an active electric line is not a good thing to do.

Many of downtown Boston's buildings have heating and even cooling systems powered by steam. The steam is carried in high-pressure pipes, which have enormous force around bends. Any change to the steam lines results in an elaborate rerouting dance with pipe supports and all types of interim stages and conditions.

Sewer lines contain the proverbial fluid that flows downhill.
Unfortunately for utility planning and construction, it doesn't flow
uphill. This means that changing the sewer line position requires
a careful study of direction and elevation to make sure the flow is
always going the right way.

To handle one utility system would be enough. To handle 30 or so different systems, all requiring staging, all interacting with each other, was a great

Wooden bracing reduces the risk of a cave-in for workers relocating a utility line under a Boston city street late at night. Utility lines like this one were in the path of the new super-highway tunnels and were relocated years before the digging of the massive tunnels began.

task. Added to that is the fact that engineers were not really sure where the utilities were, since they were built years ago. To sort all this out was a job of almost unimaginable complexity. But it was done, and amazingly, with minimal disruption.

A HIGH STAKES GAME OF "BEAT THE CLOCK"

Once the utilities were moved out of the way, it was time for the slurry walls.

Even with clear ground, slurry walls have some special construction requirements of their own. Slurry walls are constructed in panels, between six and

Opposite:
A hydraulic clam prepares to drop its load of soil into a waiting dump truck. After this load is dumped, the clam, its claws open, will be inserted back into the slurry wall trench. When it reaches the bottom, the claws will close and scoop up soil, and the process is repeated. When the trench is deep enough, it is cleaned, and reinforcement and concrete are placed in it to build the slurry wall.

20 feet long. The walls are constructed in panels because an open slurry trench is inherently unstable. The liquid slurry in the trench will keep the hole open for only so long. The trench will cave in on itself over time, so it must be filled with reinforcement and concrete as soon as possible after excavation. In that sense, slurry wall construction is like a game of Beat the Clock.

The basic method for construction involves the following steps:

PRE-EXCAVATION.

At the start, the top ten feet or so of ground is cleared and replaced with engineered backfill. This provides for better slurry-wall construction in the top layers of ground, which are usually the places where masses of buried obstructions are found.

PLACEMENT OF GUIDE WALLS.

The guide walls are five or six feet deep and are constructed on either side of the trench to be dug.

EXCAVATION.

The trench is filled with liquid slurry as it is dug.

Left: A crane prepares to drop a reinforcement
bar cage into the slurry trench.

Above: The reinforcement is gently lowered
into the trench.

"Every time I drive through Boston,
the Big Dig looks a century away from
being finished, and I can't comprehend
how it will come together."

— Mike Boylan, Hingham, MA

CLEANING.

Loose sand and clay that may have fallen into the trench during the digging is removed.

PLACEMENT OF REINFORCEMENT.

The reinforcement may be a cage of steel reinforcement bars, or it may be steel I-girder sections. A crane lifts the massive reinforcement, as much as 120 feet long, off the ground and gently positions it at the top of the trench. Then, in a few moments, the reinforcement is lowered into the trench. To the casual observer, it looks as if the earth is simply swallowing up this massive spiderweb of bars.

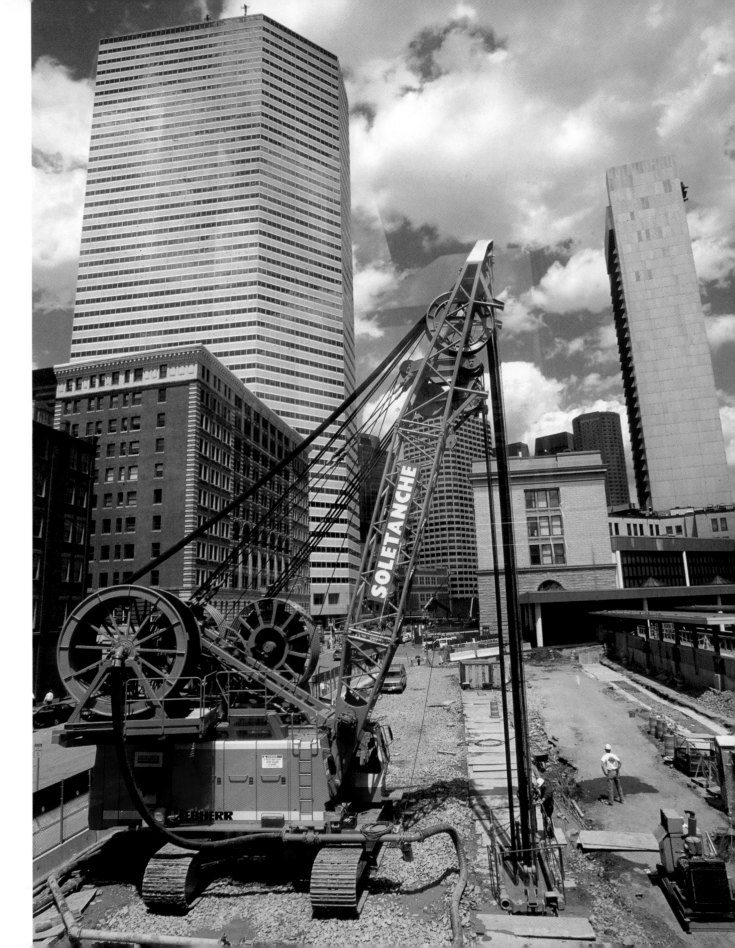

A worker in front of one of the most advanced slurry-wall machines in the world. Digging one six-foot section of slurry wall at a time along Atlantic Avenue, a team of French technicians from the Soletanche Company worked with Boston's Local 4, engineering operators to keep the rig running smoothly for nearly three years.

PLACEMENT OF CONCRETE.

With the reinforcement in the trench, concrete is pumped in.
The heavy liquid concrete flows to the bottom of the trench,
displacing the lighter liquid slurry. Eventually the concrete hardens,
and a solid wall going down 100 feet or more is complete.

This is the "ideal" description of slurry-wall construction. But the greatest
difficulty with the process is that once the initial digging of the trench is begun,
the clock starts ticking, because the trench is unstable. As in any game of Beat
the Clock, sometimes the buzzer goes off. Downtown Boston, in particular, had
some special challenges for slurry-wall construction. Large boulders left over
from glacial times buried obstacles such as parts of old wharves, seawalls and
buildings, and dozens of old underground foundations all stood in the way of
the walls.

The Central Artery was a vast laboratory for some of the most extensive
design and construction of slurry walls ever attempted. Yet this lab had one
further complication, a sort of pièce de résistance for slurry-wall construction.
About half of the walls needed to be built directly below the existing elevated
highway. The picture on page 112 shows how the reinforcement hanging from
the crane is suspended in midair. Not so for the slurry-wall panels underneath
the existing Central Artery. There was as little as 15 feet from the ground to the
beams of the highway above. How could one possibly build a slurry wall under
these conditions?

The answer was to develop elaborate methods of "low headroom" design
and construction. The methods included development of special low-headroom
cranes and rigs. Trenches were excavated using hydromills, special cutting
machines that could be slipped into the trench under these conditions. The
reinforcement required for the walls, typically steel I-sections, to be spliced into

*I can't wait until it is done, not because
of the construction hassle, but because
of the way in which it will change the
face of Boston. The future of this city is
the revitalization of the waterfront, and
the result of the Big Dig will be to unite
the old and new Bostons. Instead of
the overhead highway dividing the city,
we will have parks and recreation areas
and pedestrian walkways. Sure it is an
expensive project, but I think it is worth
every penny.*

—David Rosenthal, Boston, MA

14-foot sections so that it would all fit below the existing highway. With this carefully designed procedure, it was possible to build all of the difficult low-headroom panels under extreme conditions. When all was done, Central Artery/Tunnel constructors had completed the most difficult and extensive low-headroom slurry-wall construction ever attempted, with only rare cases of the buzzer going off when panels were kept open too long. This particular—largely unheralded—feat will be studied and applied by other tunnel builders for decades to come.

With the utilities moved and the slurry walls in place, there was still one very large obstacle in the way of the tunnel. It was the existing elevated highway itself. The old highway is supported by footings—piles and concrete pads—which are directly in the path of the digging. If the tunnel were excavated without changing the support conditions, the highway would have had to come down and traffic in the Boston area would have come to a grinding halt.

But more about that in the next chapter.

A milling machine used to excavate a slurry wall trench. This machine was specially designed to fit the "low head room" conditions for digging underneath the existing elevated highway. It was so difficult to keep maintained, workers called it the mother-in-law.

CHAPTER EIGHT

Underground in Downtown

ITH THE TED WILLIAMS TUNNEL almost finished, the business community and downtown neighborhoods braced themselves for the inevitable invasion. After five years of construction, one quarter of the Central Artery/Tunnel Project was complete. Work crews and heavy machines moved away from the harbor's edge and into the city's center. It was 1995 and the pace of work was about to double.

In downtown, utility relocation and slurry-wall construction had been under-way as early as 1989, in small isolated work zones. Now the Big Dig's yellow-and-blue construction barriers became ubiqui-tous in the center of Boston. Convoys of dump trucks, cranes, front-end loaders, backhoes, street sweepers, pump trucks, and hundreds of extra police directing

When it comes to complexity, the downtown construction takes the cake. Vehicle and pedestrian traffic and a constant demand for utilities and services leave little room for a large and complex highway project. Often work is conducted between 6:00 p.m. and 11:00 p.m. to minimize the disruptions to businesses and neighborhoods.

Keeping the roads open for residents, commuters, tourists, and travelers passing through Boston has been a major accomplishment of the Big Dig. But the signs can make people's heads spin!

Opposite:
A Boston police officer stands directly above the deepest point of the Big Dig. Without the help of police officers, pedestrians would find crossing downtown streets a major challenge.

traffic became part of the city's daily makeup. Thousands of construction workers laid siege to the city every day—and night—as a major assault gripped the city. Even after years of exhaustive preparations, no one knew what to expect.

The downtown construction area of the Big Dig is best defined as anything within 1000 feet of the Central Artery (I-93) as it runs for a mile and a half between the North End and Chinatown neighborhoods. Most of the highway tunnel under construction in this area is beneath the existing elevated highway. Along the new tunnel's path are skyscrapers, stores, restaurants, train stations,

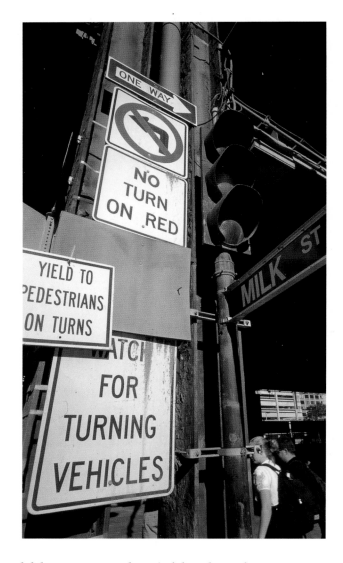

banks, luxury hotels, tourist attractions, old homes, and neighborhoods.

The new I-93 tunnel begins its path north through Boston's oldest commercial area, the Leather District. The tunnel makes a steep dive to its deepest point just outside the Leather District as it passes beneath the Red Line subway at South Station. Continuing north in front of the Federal Reserve Bank of New England, 120 feet underground, the highway veers to the right and then turns sharply to the left as the tunnel quickly climbs to its highest point. Passing over the top of the Blue Line subway and just 36 inches under State Street, the tunnel moves through the waterfront district. Still banking hard to the left, the superhighway dives again to almost 100 feet, to make an underground connection

Over 150 large cranes and hundreds of smaller cranes, bulldozers, backhoes, and other machines have made themselves at home in downtown Boston.

 Rachid Kerany came to Boston 11 years ago from Casablanca, Morocco, and drives his taxi through the Big Dig every day. "I swear to God," says Rachid, "when I pick passengers up all they ever talk about is the Big Dig! No matter where they're from—Ireland, Holland, New York—they all know about it, and they ask me questions and more questions. I worry about some of the tourists. You can see them driving their rental cars or standing on the curbs looking so lost and confused in the middle of the construction. I worry about them a lot."
—Rachid Kerany, Taxi Driver

Over 4,000 workers in the trades report to their jobs each day and night on the Big Dig. The project is a union job and workers are hired through local union halls.

with traffic on the existing Sumner Tunnel in the North End. The subterranean roadway then straightens its path and begins a steady and steep climb, one last time, surfacing onto the deck of the widest cable-stayed bridge in the world, the Charles River Bridge. Trip length: one and a half miles. Approximate travel time: two minutes.

NOT ONE HOME TAKEN OR SHAKEN

The Big Dig would never have gotten its go-ahead if Fred Salvucci had not promised zero residential destruction back in the 1970s, when support for the depression of the Central Artery was being gathered. True to its promise, the project is not taking a single home and will relocate fewer than 200 businesses to build 161 lane miles of highway through the city.

The project not only took a relatively small number of buildings; officials went further. They made sure that buildings left standing would not be dam-

Opposite:
Deep beneath the city, workers pour the roadway to the future highway tunnel. Massive cross-lot-struts, braced up against slurry walls, keep the foundations of buildings from moving as cement trucks pipe in their loads through large hoses.

aged by vibrations and settlement from deep tunnel excavation. Years before work in the downtown began, the foundations of structures along the tunnel alignment were videotaped to ensure owners that the Big Dig would be accountable if it damaged their property. A crew of experts taped the condition of the foundation, stairwells, and window frames—recording anything that was already damaged or could be damaged—to protect all parties involved, including the taxpayers. Buildings that were demolished were also videotaped so an accurate record of assets could be tracked.

THE BIG DIG'S DEEPEST TUNNEL

C11A-1 is the official contract description for the Big Dig's first downtown construction contract. The contractor is a joint venture between Perini, Kiewit and Cashman (PKC) construction firms.

At $505,000,000, PKC's contract is one of the most expensive on the Big Dig for two reasons. PKC is building the deepest section of tunnel in the project, descending 120 feet beneath the street. And it is constructing part of a new subway line called the Silver Line. Instead of tearing up the busy intersection in front of South Station twice, once for the highway and a second time for the subway, transportation officials were able to save time and money by combining the work into one megacontract.

In 2003 the airport will link up directly to the South Station transportation center via the Silver Line. Today, harried travelers must take three subway lines to get to Logan—the Blue Line, Orange Line, and Red Line. As a consequence, many people choose to travel to and from the airport by taxi or car, adding considerably to the city's traffic woes. The Big Dig's five-year acceleration of the Silver Line's opening will be an ancillary benefit for commuters and visitors.

SINKING SKYSCRAPERS!

The utility work on this contract was unusual. In some areas, utilities were moved in advance of the tunnel work under a separate contract. But PKC did their own utility work along with digging nearly a mile of slurry walls, some as deep as 145 feet. After removing the tangled web of antique utility lines, including a monster eight-foot-wide sewer line, excavation began between the two slurry walls—the east wall and west wall.

PKC began digging at the southern end of the contract area. They removed the first of the excavated material, historic fill from 150 years ago, with a forty-foot-long backhoe. As PKC worked its way north, digging deeper into Boston's blue clay, it unearthed glacial boulders and bedrock every foot of the way. By the spring of 1997, the tunnel was too deep for the extra-long backhoe. Crews switched to clamshell buckets lowered by cranes into the work area. The digging was slow, because the buckets had to travel through a maze of massive steel beams, called cross-lot struts, necessary to support slurry walls.

Every 12 feet that PKC dug down, they were required to place a long strut running perpendicular from the east slurry wall and west slurry wall—dig 12 feet, place a strut, dig 12 more feet down, and place another strut. The weight of the earth and pressure from the foundations of the large buildings along the way exert a tremendous force against the deep underground walls. Powerful jacks wedged the struts into place, matching the pressure from the backside of

Opposite:
The Big Dig's biggest crane, working here on Atlantic Avenue, can lift nearly 400 tons. The boom is over 300 feet long and the smaller jib-boom is 128 feet in length. The wide-angle lens of the camera makes the brick building look as though it is leaning. Happily, it's not.

the walls and creating a balanced force from either side of the slurry walls. If not for the struts, the walls would bend and collapse. The struts on the Big Dig must also carry the weight of thousands of tons of thick concrete planks that allow 220-ton cranes, city buses, trucks, taxis, and pedestrians to move about on the street above.

Another problem is water. It would be easier to cure the common cold than to keep a Big Dig construction site dry. The water table in Boston is about 15 feet below the surface of the ground, guaranteeing that tunnel construction sites are inundated with groundwater. Even though the slurry walls were sealed with grout and covered with waterproof mats, moisture still seeps in. Pumps run 24 hours a day to dewater the underground work zones.

Conditions like this call for drastic measures. Big Dig officials embarked on a dewatering campaign in 1996 to drain groundwater out of the construction area under Atlantic Avenue. Starting in the summer of 1996, PKC began pumping out groundwater and continued for nine months, lowering the water table in the area by 45 feet. This helped with the water problem inside the tunnels, but it created a sensation of sorts above ground.

BIG DIG SINKING DOWNTOWN SKYSCRAPER was a headline on the front page of *The Boston Herald* on Thursday, April 30, 1998. With all the dewatering going on, the buildings above the tunnel were settling a little deeper into the ground. After the *Herald's* story, a gaggle of television reporters with news cameras crowded into the downtown intersection of Summer Street and Atlantic Avenue. Any passing pedestrians who worked in the affected buildings were asked their opinions of "skyscrapers falling and rising." Nancy Powell, a legal secretary on the thirty-ninth floor of One Financial Center, spoke to the Associated Press. "Some people are very nervous about it. For others it's just a curiosity. The joke is, if you work on the twentieth floor, now you work on the nineteenth." It's just as well that Ms. Powell was probably unaware that 23 feet

Opposite:
One hundred feet below Atlantic Avenue, workers begin to remove the cross-lot struts, which are temporary braces. The worker flashes a light onto a vertical soldier-pile in a slurry wall.

Opposite:
The Big Dig has an excellent safety record. Workers from different construction sites compete against one another for the best safety records. Winners walk away with cash rewards.

outside her building's front door, Big Dig workers had dug 73 feet beneath her building's 47-foot-deep foundation.

Yes, the skyscrapers were sinking, but only a fraction of an inch. As water is pulled from under a building's foundation, the soil it sits on becomes compacted, and the structure sometimes settles less than a quarter of an inch. For Big Dig engineers, this is an everyday occurrence. Detailed explanations had been passed on to building owners years before the news teams showed up. In fact, property owners were not unduly alarmed. A Federal Reserve spokesman said to the *Herald* a week later, "Whatever movement there was, it was acceptable and no cause for alarm. They have instruments in place that are clearly providing this information."

Instruments galore is more like it. Over 25,000 devices are spread over eight miles of Big Dig construction, gathering readings on ground movement, vibrations, and building settlement. Seismographs, mechanical-strain gauges, vertical inclinometers, and electro-optical meters are placed in the soil, on sidewalks, inside buildings and construction sites, and anywhere else that changing ground conditions are an issue. They detect up-and-down and side-to-side movements in the building's foundation and in the ground. Call it Prevention of Angular Distortion.

Angular distortion is a major concern of construction managers. It refers to non-uniform settlement due to groundwater fluctuation. Settlement is fine as long as it's not uneven. When a building rises and falls with the water table and one side or part of the building doesn't move along with the rest of the structure, fractures in the foundation begin to appear. Fortunately this was not the case on Atlantic Avenue and Summer Street.

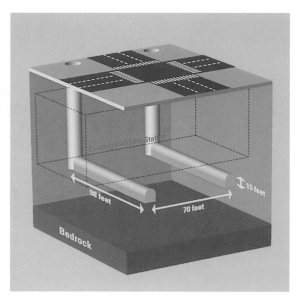

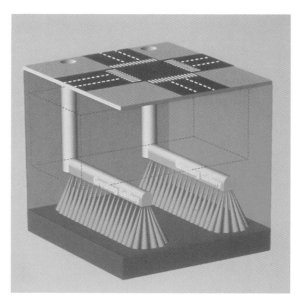

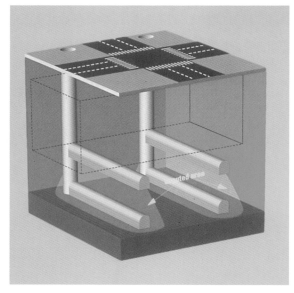

120-feet below the busiest pedestrian intersection in the city of Boston (Atlantic Avenue and Summer Street), workers built a bridge for the Red Line subway to pass over four lanes of superhighway. First, two vertical access shafts and two grouting galleries were built directly beneath the subway. Grout was injected into the soil from the grouting galleries to stabilize the ground so that three more tunnels beneath each gallery could be dug.

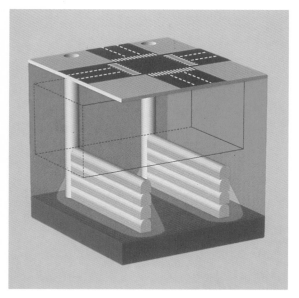

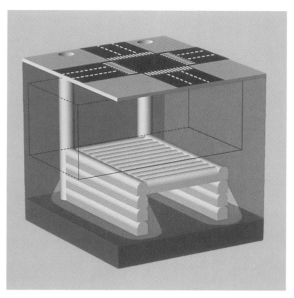

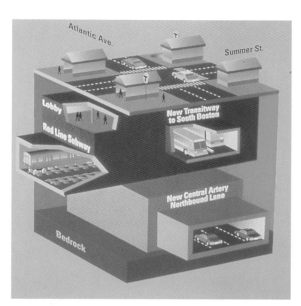

The deck of the subway bridge and/or the roof of the highway tunnel was built by digging 11 tunnels from one grouting gallery to another. When it is complete, four lanes of highway will travel beneath the Red Line subway, which travels beneath the Silver Line Transitway. The Transitway will be below the subway lobby.

BUILDING A BRIDGE UNDERGROUND

It is at Atlantic and Summer, beneath the city's busiest pedestrian intersection and most traveled subway—120 feet below the sinking skyscrapers—that four lanes of Interstate I-93 reach the deepest point in the Big Dig. As Lyndsey Layton, former reporter for *The Patriot Ledger*, explained, "It's one of the most complicated and dramatic technical tasks of the complex Big Dig. An engineering opera is taking place in dark cold tunnels, carved by hand from wet clay and rocky dirt where workers carrying oxygen are building part of the new underground Central Artery."

It's no surprise to passing motorists that about 150 large cranes are in use on any given day.

But before digging could begin, the structures above needed to be reliably supported. The underpinning of the old Red Line subway began beneath South Station's front door in 1996. It is a very important and dangerous piece of work, one of the largest and most costly underpinnings of a subway system ever undertaken. It is a microcosm of the entire Big Dig: building a superhighway beneath two functioning, crowded subway lines and the lobby of a busy transit station.

Project officials did not have a choice in the matter. The only place in the city for the new northbound lanes of I-93 was under Atlantic Avenue. That's where the Big Dig must build its highway and a section of the Silver Line subway tunnel called the Transitway.

The Transitway will carry 60-foot-long articulated buses, "the most technically advanced buses in the country," according to the Transit Authority. They will run through tunnels on electric power or natural gas and switch to diesel when they surface from the Transitway tunnels and drive about the city streets.

The Transitway placement above the 84-year-old Red Line forced the construction of the superhighway to be under the old subway tunnel. The new Silver Line is only 36 inches above the Red Line, and the station that services both the Red Line and the Silver Line sits on top of the roof of the transit line.

Opposite:
Tunnel work is
dangerous enough
but when you're
lifting 40-ton struts
and roof girders it
gets even riskier.
The tread of a special
low clearance crane
used to lift the
massive beams can
be seen between
the two workers.

During the underpinning operation, 225 tunnel workers, called sandhogs, piled into cages, six at a time, and were lowered down a 120-foot access shaft each day. The Local 88 workers could clearly hear the screeching of the Red Line subway's brakes as the trains came to a stop directly over their heads. Leo Antonelli, a sandhog who has worked some of Boston's deepest tunnels told *The Patriot Ledger*, "It's different. You can hear the wheels. It doesn't scare you, but it makes you think."

Building a bridge for the Red Line subway to pass safely over four lanes of the interstate is the underpinning's main objective. A series of 19 machine- and hand-mined shafts, each 15 feet high and 98 feet long, make up the underground bridge. To stabilize the Red Line, two tunnels were mined, 75 feet away from each other and perpendicular to the Red Line but just 36 inches beneath the old tunnel's foundation. Grout was injected from these two mined areas, 60-feet into the virgin ground beneath them. The grout material filled the earth below, stabilizing the weak soil beneath the first two shafts and creating "grouting galleries." Three more tunnels under each of the grouting galleries were safely and quickly mined. PKC now had two sets of four shafts directly on top of one another, for a total of eight. Each stack of four shafts makes up a wall for the northbound highway tunnel passing under the Red Line subway.

Once the eight shafts were completed, 11 more were mined from one grouting gallery to the other, creating the highway tunnel's roof. The entire lot, 19 mined shafts, was filled with post-tensioned cables and reinforced concrete. The steel and concrete inside the shafts give strength to the bridge so heavy subway trains can pass over the highway tunnel.

After the solid walls and roof of the highway tunnel were built, crews began to excavate the earth from under and between the shafts. To clear a path for the future highway, workers simply removed the material under the roof and between the walls of the 19 tunnels.

One night in December, 1999, at about 1:00 a.m., shortly after the last of the material under the Red Line was removed, Frank Nee, chief construction manager of the "nightcats," admired his team's handiwork. Standing 120 feet underground, Nee said, "We are creating real estate down here. This is an historical structure that no one will ever see. It's a beautiful underground cathedral with a 50-foot ceiling, buried deep under Atlantic Avenue. We're proud of it."

His pride is understandable. The operation ran 24 hours a day, seven days a week for several years, and not once did construction stop the Red Line trains from running, nor, in spite of dire predictions, did it sink any sky-scrapers. It was a major engineering triumph.

THE SAME BUT DIFFERENT

Farther north, the tunnel's construction is the same in concept but very different in its method. In front of South Station, the deep tunnels have been built in the traditional cut-and-cover style. The contractor cuts a swath as wide and deep as the tunnel to be built, builds the tunnel, covers it over with back-fill, then rebuilds the city street that was ripped up years earlier. Because the highway tunnel is built from the bottom of the excavated area upward toward street level, this technique is also called bottom-up construction.

North of South Station, where the tunnel is not so deep, contractors have been able to take advantage of top-down construction. The work is car-ried out between the slurry walls from the surface and moves downward, building the roof first and ending with the floors.

As soon as the roof is installed, heavy machines, bulldozers, and other excavators operate in a 10-foot shallow tunnel, severely restricted by the street decking above. Earth removed from the tunnel must be taken out through small holes in the roof deck, called glory holes, an old term from the gold-rush mining operations. It's a difficult, confining operation. However, the difference for the neighbors living every day with the Big Dig is significant.

With the roof put into place first, the city street above returns to normal while heavy machines and work crews build the tunnel out of sight and, one hopes, out of mind. It is also less expensive. Top-down construction uses fewer cross-lot struts because the permanent roof girders eliminate the need for an entire row of temporary beams.

THE GREEN MONSTER LOOMS

The biggest difference between the downtown construction work and other Big Dig contracts is the Green Monster overhead. Over 190,000 vehicles pass along it every day with ten hours of gridlock traffic. The weight of the original roadway—93,000 tons of steel and 459,000 tons of concrete—the load of a couple of hundred thousand trucks and cars each day, and thousands of tons of temporary steel support beams makes creating the underpinning of this old highway a high-wire act. There is not another city or country on the globe that has attempted to build a massive subterranean superhighway below while supporting an old, crumbling but still in use structure above.

The only way to clear a path for the enormous tunnel beneath the old highway was to remove every one of its original foundations. The Big Dig has cut out hundreds of steel and concrete columns embedded 60 feet into the ground, and replaced them with temporary columns placed on top of slurry walls.

"Normally, underpinning is avoided and used as a last resort. Even then, only a small portion of a roadway is underpinned—a column here or there when conducting repair work on an elevated roadway," explains Bill Goodrich, one of the lead engineers for the underpinning operations. "On the Big Dig, we removed the support structure of over a mile of interstate highway without ever stopping the flow of traffic above or below. Not one piece of the old road ever moved, it's exactly where it was in the 1950s. It was exciting to see so many people's hard work pay off."

Approximately 33,000 tons of temporary steel is used to prop up the Central Artery. Flat jacks, a jacking system developed by Modern Continental and Weidlinger Associates, were employed. The jacks were put in place under temporary support beams, which surround the old columns destined to be removed. The flat jacks were then injected with oils, exerting an enormous amount of pressure between the old highway and the temporary beams that eventually took the full load of the existing structure. The weight from the old columns was slowly transferred to the new interim supports, many flat jacks working together with electronic coordination to make certain the jacking was done evenly. This part of the underpinning operation took place during the peak hours of traffic.

The activation of the flat jacks exerted a tremendous force onto a 40-ton grade girder, part of the current support structure for the Central Artery, bending it one half inch. This prepared the underpinning system for the full load of the highway. An epoxy was then injected into the flat jacks, forcing the oils out of the system and leaving behind a hardened mass. The flat jacks, full of epoxy and applying pressure to the old highway, become part of the underpinning system, literally molding themselves into the supports. They are still there today, left between the beams and girders holding up the highway. They will be removed with the rest of the elevated highway when it's finally scrapped.

With the load of the highway transferred from old to new supports, the cutting out and removal of the 1950s columns began. This part of the operation was executed between 1:00 a.m. and 5:00 a.m. to reduce the chance of a catastrophic failure. The work took on a kind of mythic, otherworldly feel. For the first cutting, near Quincy Market at State Street, the engineers and

Opposite:
An articulated dump truck removes dirt from a downtown construction site as a bulldozer continues to excavate beneath cross-lot struts and the original elevated highway.

contractors solemnly gathered on October 5, 1997. It was a mild, pleasant, clear night. Traffic maneuvers were scheduled to begin at about 11:00 p.m., when the flow of cars and trucks thinned enough on top to permit shutting down lanes. There was a nervous chatter and excitement in the group as the appointed hour approached. The city quieted, and it was time to begin. The acetylene cutting torches were lit. The engineers watched with a combination of pride and morbid fascination. It's one thing to prepare engineering calculations on paper. It's quite another to watch the existing support of the bridge, a column that faithfully held its load for 50 years, being burned away. Would the engineering numbers really add up?

The torch cut through the steel in a "nibble" pattern, taking a little bit from this side, a little bit from that side. After 20 minutes or so, it was time for the final cut. In a blaze of sparks, the last piece of old column was torched through. Nothing moved—the new grade beams and supports held the weight exactly as they were supposed to. Load transfer for the first column was complete. Of course there hadn't been doubt in anyone's mind. But there were a lot of sighs of relief.

Today thousands of workers toil away beneath the underpinned interstate, while construction equipment, buses, trucks, cars, and pedestrians pass under and over the roadway. This fulfills another promise—that the interstate would remain open throughout construction.

A few feet beneath pedestrians, taxis, buses, trucks, and cars a Big Dig worker builds temporary supports to hold up temporary roadways. About one third of the project's budget pays for mitigation efforts to keep traffic moving during construction.

WHO NEEDS PHONES?

Losing utility service because of the Big Dig was the number one concern of businesses when the Boston Chamber of Commerce took a poll of downtown companies before the Big Dig began. Members of the chamber were surprised. They were certain the biggest concern would be traffic, which came in at number two.

Ironically, it was years after the nearly flawless relocation of thousands of miles of utility lines carrying over 25,000,000 phone calls a day that the Big Dig had one of its worst utility blunders. Eleven hundred phone and data transmission lines were drowned in a blast of water from a water pipe that Big Dig crews had relocated under Atlantic Avenue on a Friday afternoon in September, 1998. By 9:30 the next night, unbeknown to anyone, 15 feet of water had collected, forming an underground moat around the Boston Harbor Hotel and the Rowes Wharf office and condominium buildings. Not until a torrent of water made a direct hit into the luxury complex's phone room was anyone alerted. Within minutes a tide traveling through the ceiling of the parking garage dumped thousands of gallons of water, destroying over $250,000 worth of electronic phone gear. Hotel guests and over 40 businesses were without phone service for nearly a week. Big Dig engineers and construction managers surrendered their project cell phones to the guests, businesses, and residents of the complex. Bell Atlantic technicians worked around the clock until service was restored. It was not a happy moment.

THE BELLY OF THE BEAST

Jack Robertson is the resident engineer for Big Dig contract C15A-1. His 1400-foot-long work zone stretches beneath the widest section of the old elevated Central Artery. The mayor's office is just across the way, and Boston's most vocal and watchful neighborhood—the North End—surrounds one half of his contract. His construction field office is in the administration building for the Sumner and

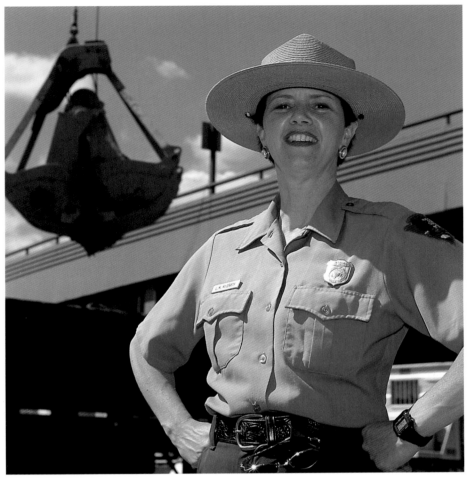

Life goes on beneath the elevated highway. National Park Ranger Lauren McGrath conducts tours along Boston's Freedom Trail for the National Park Service. The Freedom Trail passes directly under the Central Artery as it takes tourists from downtown Boston to the North End. "People from all over the world have heard about the Big Dig, so when I get under the artery I say, 'This is the Big Dig! When it's done this ugly green thing is going to be gone and it's going to be great!'"
—Lauren McGrath
National Park Ranger

Callahan tunnels, literally sitting on an island between the entrances and exits of the two tunnels. The northern end of his work zone is near the Fleet Center, Boston's indoor sports and entertainment mecca.

One of New England's most famous tourist attractions, the Freedom Trail, and 1,500,000 tourists blaze through the C15A-1 construction site each year. If Jack's construction crews delay traffic coming out of the Sumner Tunnel, airport officials call him and complain. If work crews delay traffic entering the Callahan Tunnel, the downtown traffic stops in every direction. One night two cars plowed through two of his construction barriers, plunging forty feet into his construction site, landing one on top of the other. Work crews that night avoided injuries, but one driver was killed and the other was lucky to have survived. Things are not easy for Jack.

"This is the only construction site in America where a driver enters the city from an international airport and passes through city streets, traffic lights, and pushcarts selling bananas before they can enter onto the interstate," laments Robertson. "We are trying to build the widest section of tunnel through the most demanding part of the city. I am dealing with interstate traffic, local businesses, city hall, residents, tourism, and an international airport. If we get the residents happy we're bound to upset the businesses. There's not a day that we're not bothering someone."

Robertson, however, is building one of the most interesting pieces of underground highway on the Big Dig. His tunnel, 1400 feet long and 350 feet wide, is a $455,000,000 endeavor for the joint venture of J. F. White, Slattery, and Perini. Their mission is to dig down 115 feet and build eight lanes of highway, replacing the six lanes of traffic on I-93. Between the eight lanes of deep highway and the busy North End streets, the contractor is assembling one of the world's first high-speed underground interchanges.

In the future, traffic from Logan International Airport, coming into Boston through the Sumner Tunnel, will exit onto I-93 northbound through a series of complex underground ramps. The ramps are four lanes wide, with more capacity than the Central Artery's present north- or southbound lanes. The contract is only 1400 feet long but demands the construction of over 6000 feet of ramps alone. To make a complicated situation even more so, each underground ramp is at a different elevation, with sharp banked curves, and must be wedged between the eight lanes of superhighway. And as elsewhere, the old elevated highway above this construction site needs to be underpinned. Its 1950s columns have all been removed, and new temporary supports currently

> "The Big Dig is a major headache. The promise is that it will improve traffic as well as make the landscape of Boston nicer, but it's been going on for so long. And it's so expensive. I'm sure Boston will be a nicer city when it's done, but I could be a grandmother by then— and my kids are still in grade school!"
> —Marianne McGee, Bridgewater, MA

hold it up. The interstate remains open while Jack and his crew tunnel deep in the ground below and build their maze of ramps and superhighway.

FIRST DIG TO BIG DIG

The Big Dig has uncovered some of New England's most historically significant discoveries as it plows its way through history. And as it turns out, Jack Robertson's contract has uncovered the mother lode of colonial artifacts.

The gargantuan highway tunnel swallows two blocks of American history as it passes through Boston's first Dig, the Mill Pond, and a forgotten area called Paddy's Alley.

Jack Robertson remembers, "I got a call on my radio back in November [1999] that the contractor had hit some big stones and I should page the project's archaeologist ASAP." A large Caterpillar bulldozer, operating about 15 feet underground, had unearthed two huge, 2200-pound millstones from the 1700s.

Bob Hasenstab, Big Dig archaeologist, recalls, "When I arrived, one of the guys had loaded a millstone into the back of his pickup truck and was planning on taking it home to his backyard. I didn't blame him for wanting it. They are in perfect condition, preserved for hundreds of years by the mud. [Mill Pond was filled in around 1808.] Usually contractors get to keep what they dig up, but everything on this project belongs to the state."

But what really excites the Big Dig's archaeologist is the old creek's muddy bottom. Hasenstab explains to a *Boston Globe* reporter, "We have samples of the bottom of the canal, and we're going to be looking for pollen and insects and parasites." This, and other finds on the Big Dig, are giving archaeologists a chance to look at Boston's past with a fresh environmental perspective.

A few hundred years ago, just down the road from the mill, was Paddy's Alley, a narrow neck of land connecting the North End with Boston proper. It was a busy part of the city from the 1600s up until the construction of the John F. Fitzgerald

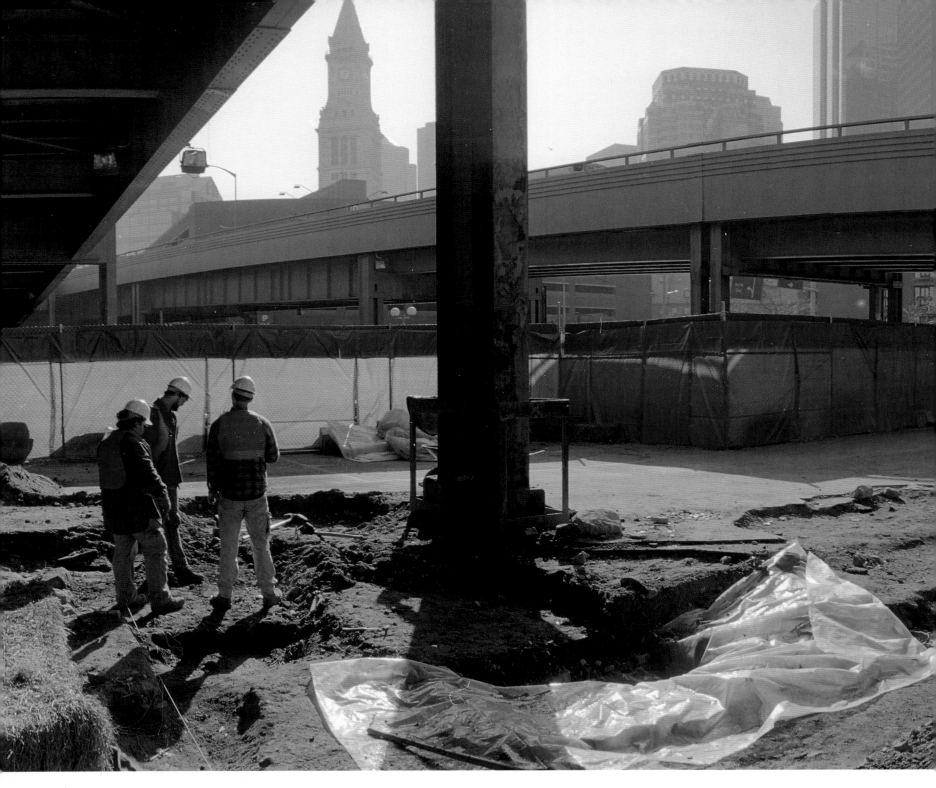

Archaeologists began excavating Paddy's Alley near Boston's North End in the fall of 1992. When this picture was taken, they had no idea they would unearth so many significant finds from Colonial America.

Katherine Nanny Naylor's privy and the remains of John Carnes's pewter-smith operation were found here. Ironically, the elevated highway preserved the site from the more invasive construction of office towers.

An archaeologist on Spectacle Island uncovers arrowheads and other finds during the summer of 1992. The shoe and porcelain were typical of finds in Paddy's Alley and other parts of downtown.

ARCHAEOLOGICAL FINDS:
FROM MUSKET BALLS TO A BOWLING BALL

"We are very lucky to have found the likes of Paddy's Alley," explains Big Dig archaeologist Paul Mohler. "Boston has seen a lot of construction and disruption over the last 370 years, yet these sites were relatively undisturbed. The credit belongs to Boston University for their outstanding and extensive research early on. They pulled old maps of utility lines, buildings, roads, and subways, looking for ways to excavate with minimal disturbance." After locating Paddy's Alley with ground-penetrating X-ray equipment, archaeologists began the process of recovering long-hidden treasures.

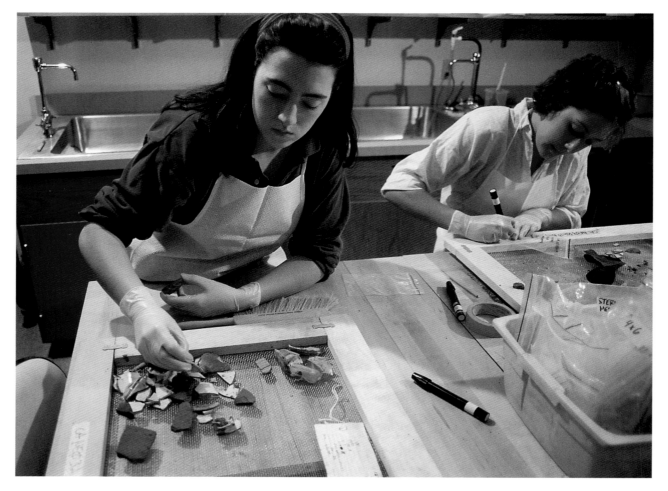

In a Boston University laboratory archaeologists preserve artifacts uncovered from the Big Dig sites. The Commonwealth of Massachusetts owns everything of historical significance recovered from sites. Over 200,000 items including tools, shoes, pipes, arrow-heads, musket balls, toys, bottles, and even America's oldest bowling ball have been uncovered by archaeologists and Big Dig workers.

Expressway in 1954, as the first phase of the raised highway was named. Until then, the area had bustled with activity from homes, stores, and warehouses. Discovered here, beneath a paved parking lot under the elevated highway, was debris from John Carnes's successful pewter shop, including wine-bottle seals embossed with his name. John was apparently a citizen in good standing in the mid-1700s, having been a lieutenant colonel in the Ancient and Honorable Artillery Company, drinking expensive imported wines and living the good life.

In nearly the same location, a hundred years earlier, Katherine Nanny Naylor was living a different sort of life, at least according to her toilet. In 1708 someone sealed Katherine's privy, or outhouse, with an impermeable layer of Boston's blue clay, preserving its contents remarkably well. As it turned out, Katherine used the outhouse as a garbage dump as well as for its more usual purpose. Reaching deep into her past, archaeologists pulled out her diary, wool fabrics with expensive weaves, 158 scraps of red silk, scissors, 53 bags of cherry plum pits, a pipe stem, five shoes, and the oldest bowling ball in America!

It appears that in 1663 Katherine Nanny's first husband died and left her fairly well off, with a son, a daughter, and a house. But things took a turn for the worse when she married Edward Naylor and had two more daughters. In 1671 she asked for a divorce, accusing him of beating her, kicking one of the girls down the stairs, and "abuses of the marriage bed." To top it all off, his mistress tried to poison her. Katherine received one of the first legal divorces in the New World, and Edward, the sot, fled to New Hampshire with their pregnant servant girl. He wrote home to Katherine begging for his clothes, saying, "I have none to wear, especially linen and shoes." But it appears she tossed them in the privy.

The cherry pits, however, have stumped archaeologists. She either ran a successful cottage industry baking cherry pies or was making a popular colonial alcoholic drink called cherry bounce. Whichever, Katherine did live to the ripe old age of 86.

FINISHING A TUNNEL

Mike Ryan, the engineer in charge of decorating the I-93 tunnel, tells it like it is. "I completed the tunnel finishes for the entire Ted Williams Tunnel, and now I am finishing the whole downtown tunnel. I'm in charge of everything you'll see through the windshield of your car as you pass under the city. That's 800,000 feet of architectural accoutrements." The translation: Mike is installing miles and miles of tiles, stainless steel railing, pavement, wall panels, ceiling panels, fire protection, utility conduits, and a lot more. Duration: six years. Cost: $163,561,844.

When several construction contracts are completed and the tunnels are mostly built, the likes of Mike Ryan come along and perform follow-on work. Mike knits the completed highway tunnels together. Seven large construction contracts that comprise the entire downtown tunnel become Mike's single contract for tunnel finishes. When he is done, you'll never know where one contract ended and the next began. Mike is going to be very busy hanging his architectural accoutrements from Chinatown to the old North End.

The Boston Fire Department requires a two-hour fire-protection coating in the ceiling, costing $20,000,000, along with five miles of stainless steel handrails and 50 stainless steel doors for emergency exits. Over 120,000,000 eight-inch tiles and millions of smaller ones cover the walls, while four-inch-thick concrete ceiling panels must be hung from the roof girders. On the road surface, Mike and his crews apply tons of concrete and asphalt three inches thick throughout the tunnel. And they paint the town white and yellow with over 23 miles of reflective lines.

Perhaps Mike's last job will be to rehang the old state highway sign that was taken down during Big Dig construction and put away for safekeeping. It reads THE JOHN F. FITZGERALD EXPRESSWAY. Wouldn't our thirty-fifth President's grandfather be impressed?

The Mother of All Interchanges

HEN "STORMIN' NORMAN" SCHWARZKOPF needed someone to back him up in the Gulf War, it was General William F. Flynn that he turned to. Flynn, a three-star general, moved an army of U.S. troops and M-1 tanks from Germany to the Saudi Arabian desert to fight the war Saddam Hussein called "the mother of all wars." In 1992, after deploying the largest collection of military hardware and manpower since World War II, General Flynn retired—and became the deputy project director of the Big Dig. "Mobilizing the most powerful army in the world wasn't easy. I really thought working on this project was going to be a piece of cake, but baby was I ever wrong. The Big Dig makes army life look pretty

Six levels of superhighway tunnels, surface roads, and viaducts will weave their way through 60 acres of railroad lines and swampy water in the heart of Boston. For nearly two hundred years the South Bay has been a crossroads between downtown and South Boston's industrial facilities. Now, at this point, the Big Dig is reconstructing a complex interchange of two intersecting interstates, I-90 and I-93.

Before: I-93 north- and southbound traffic passes through the picture from left to right. I-90, the country's longest interstate at 3,081 miles, extends from Seattle, Washington, to a dead-end in the South Bay Interchange.

It's no surprise that General Flynn calls the Big Dig's most complex interchange "the mother of all interchanges." It's a 60-acre site where two interstates come together to form a mass of ramps, tunnels, surface roads, and boat sections (belowground open-roof roadways). The interchange, also known as the South Bay Interchange, places I-93 up and over I-90, and allows 28 separate routes to connect to local roads or other destinations. Every day 275,000 vehicles and 290 Amtrak and commuter trains roll through the South Bay Interchange while an army of construction workers and equipment rebuilds it.

South Bay has been a busy intersection of industrial traffic for over 200

years, whether it's been ships, barges, trains, trucks, buses, or cars traveling through the area. The earliest bridge was constructed across the bay when it still was a bay, and connected Boston with Dorchester. When the Central Artery and Southeast Expressway cut through these historic crossroads, it severed many local roads that provided connections to and from the heart of Boston. The Big Dig has already reknit some of the cross streets and will do even more to repair the damage from the 1950s expressway.

The work is dangerous and not for the weak. "We're demolishing an existing bridge and erecting the segments over electrified rail," explains John

After: The mother of all interchanges will incorporate 29 lanes of traffic between I-90 and I-93 and several local roads. Two levels of tunnels and three levels of viaducts divided by a surface road create a virtual slice of highway cake.

Foster of Modern Continental, in *Engineering News-Record*. "We're also building just about every type of bridge in the industry just to maintain traffic. There are five temporary steel-stringer bridges, three temporary modular-panel bridges, two permanent concrete bulb T-girder bridges, and a permanent box-beam bridge plus substructures." One of the temporary bridges caught fire when a train's locomotive, waiting to pull into South Station, was parked under the structure. The engine's hot exhaust ignited flammable material beneath the bridge. Fortunately there were no injuries.

Traffic headaches in the area have been lessened because of 30 temporary ramps and 13 temporary viaducts (roadways supported by pillars). The largest temporary ramp carries the entire southbound side of I-93, and is known as IVAS, the Interim Viaduct over Albany Street. One summer weekend in 1997, General Flynn directed "the mother of all traffic moves" from the Big Dig's "traffic war room" overlooking the interchange. With the assistance of police, Big Dig crews detoured 70,000 southbound cars a day from the 50-year-old viaduct onto IVAS, which is expected to be in place for five years. Demolition of the old southbound lanes began the next day.

When complete, the South Bay Interchange will be devoid of office buildings, parks, or any kind of pedestrian access. Its sole function is to keep hundreds of thousands of vehicles a day moving straight through a 60-acre chunk of the city. The vehicles will pass over, under, and alongside one another, traveling on six levels: three tiers of viaducts stacked on top of one another and two levels of tunnels, with surface traffic between. Not exactly glamorous, but certainly critical to the overall success of the Big Dig.

The tangled maze is so complex that the Big Dig hired six engineering firms and 14 general contractors to design and build the interchange, along with hundreds of subcontractors and union workers.

THE CRANE FARM

At the peak of the Big Dig's construction, during the summer of 2000, times have never been more demanding for Local 4, the International Union of Operating Engineers. "We have a lot of work on world-class technology at play on the Central Artery. It's great for our members, but we're stretched thin, with over 900 of our 5000 members working on the project. Keeping the Big Dig staffed with licensed engineers has taxed the union's resources," says Bill Ryan, Local 4's head manager. "I have earthmovers, but I need crane operators. I'm working my crane network with our sister locals from around the country and pulling operators up from the South, out west, and all over New England. The state has had to step in and expedite licensing hearings so we can get operators in the seats over at the Big Dig."

Along with the diversity of technology is the diversity of equipment. "We have all the construction toys in the world here," claims Mike Bertoulin, area manager for the South Bay Interchange. There are pile drivers, drill-shaft rigs, deep-soil-mix rigs, slurry buckets, jet grout rigs, cranes, loaders, excavators, including the world's largest. But what really dominate the skyline are the tall booms of the cranes, their long necks swinging back and forth as they build the interchange.

The Crane Farm has become the South Bay Interchange's nickname. Roughly a third of the entire project's tall cranes are in this area. There are so many cranes, so close to one another, operators have to keep tabs on each other's crane booms to avoid disastrous collisions. The large rigs require a crew of two and rent for $105,000 a month. Try that for a car payment.

Tagmen use radios, horns, and over 16 hand signals—their own sign language—to coordinate "picks" with the operators. Sometimes two cranes lift a single load together in a delicate and dangerous job. Computers monitor the

weight and the degree of pitch to the boom for the operator to ensure safety, which is on everyone's mind.

Crane operators are a breed apart. They are highly visible and ultimately responsible for the lives of every tradesman who works beneath their booms. If an operator doesn't instill confidence in the workers below him, they will refuse to work under his crane.

In an article in *The Boston Globe* in September 1998, Crane Farm operator

Bill Smith explained, "I'd be lying if I said I didn't worry. There's a lot of pressure. When we pick something up, it's usually going over someone's head and into a small place." Mike Ryder, operating a massive Manitowoc 4100, echoed Smith's concerns. "You want the people on the ground to go home with everything they came here with—with all their fingers."

In October 1999, early one cold and wet morning at the southern end of the massive interchange, disaster nearly struck. During the height of rush-hour traffic, at about 8:15 a.m., the boom of a Big Dig crane started to fall apart. The operator heard the ominous sound of metal giving way. He had to make the decision every operator hopes to avoid: "Where do I lay this boom down in a hurry so I don't kill anyone?"

If he swung the six-story-long boom to his right, he would have laid it across the congested northbound side of I-93. If he swung it too far to the left, he would have dropped the eight-ton boom into the intersection of two crowded local roads. Somehow the operator rotated the boom to just the right spot before it came crashing down inside the construction site. The only injury reported was one carpenter's broken thumb—a reminder that the Crane Farm is a dangerous and crowded work zone.

THE MOST MASS EVER JACKED

Towering above the rest of the Crane Farm are three bright yellow tower cranes shipped to Boston from the Netherlands. Each crane represents one of three separate tunnel-jacking efforts that together make up the largest comprehensive tunnel-jacking operation in the world. There have been jacking projects that included utilities, pedestrian walkways, and small roads. But never has a nation jacked such a collection of superhighways beneath a very busy railway.

Tunnel jacking is the pushing of any premanufactured object, usually a pipe or sewer, through the ground and beneath an obstacle. Utility lines are

Opposite: Pete Bergnazzani climbs 191 feet every morning to reach the controls of the Big Dig's largest free-standing tower crane. Boston's Local 4, the International Union of Operating Engineers, provides the Big Dig with all its licensed crane and machine operators. It takes four years of apprenticeship to be licensed on the largest cranes.

jacked under busy highways to prevent traffic disruptions. Small local roads and pedestrian walkways are jacked under railroad beds, avoiding costly and expensive disruptions to rail service. Highways, however, are rarely jacked.

Why jack one now? Because this tunnel will pass directly under the tracks of South Station, where 290 trains travel in and out every day. Commuter and Amtrak trains depend on a system of interconnected rail lines that cannot be separated from one another, thereby eliminating the option of taking out one track at a time and building a cut-and-cover tunnel. Since the Big Dig's early days, ten years ago, officials have been encouraging commuters to leave their cars at home and take mass transit. To tear up the city's mass transit at the same time the roads are under construction would be a public relations—and traffic—disaster.

Before the Big Dig, the largest objects ever jacked in the United States were 30-foot-wide culverts used to capture floodwaters in California. An 11-foot drainage line was jacked under the Davison Freeway in Michigan. In the South Bay Interchange, three 78-foot tunnels are being jacked as part of one contract.

"This is the largest jacking attempt ever made when you add up the height, width, and lengths of the three tunnels. There have been longer, wider, and higher jacked tunnels but never has so much tunnel mass been pushed as a part of one project. It's the biggest!" proclaims Jason Rodwell, a tunnel-jacking engineer from London.

JACK-IT-YOURSELF

Imagine you have one weekend to jack your house from one side of the street to the other. If this were possible and Home Depot had a do-it-yourself tunnel-jacking kit, the instructions would read: (1) Dig a hole bigger than your house. (2) Put your house into the hole with the front door facing the direction you want it to be jacked. (3) Freeze the ground under the street. (4) When ground

The Big Dig employed traditional tunnel-jacking methods to push utility lines from one side of busy highways and city streets to the other. Left: Dangerous work, as Big Dig tunnel workers relocate utility lines in downtown's weak and soft soil. Local 88, tunnel workers proudly call themselves "sandhogs" and are often second- and third-generation miners. Below: Inside a small jacking pit in downtown Boston, sandhogs prepare to unload a cart of material removed from a tunnel operation. After sandhogs dig and clear a cavity in front of the tunnel sections being jacked, the two large hydraulic jacks on either side of the workers push the next section of tunnel out of the jacking pit, extending the tunnel by the length of another section.

Ramp-D is the smallest of the three tunnels jacked, but if it's the baby

under street becomes a block of ice, start chipping it away from your front door. (5) Remove frozen material from front of house, taking it through the house and out the back door. (6) Attach 50 Home Depot jacks to the back of your house. (7) Jack house forward three feet. (8) Repeat step five. (9) Jack the house three more feet. (10) Jack house to desired location, informing neighbors of new address.

The tunnel-jacking operation is part of a contract won by the joint venture of Slattery, J. F. White, Interbeton, and Perini. Using slurry walls, Slattery and company first built three jacking pits. Each pit is nearly 60 feet deep, deeper than the foundation of some skyscrapers. After crews excavated the earth from the jacking pits, they began casting the three highway tunnels. While crews built the highway tunnel sections inside the jacking pits, other teams of workers began to freeze the soil under the railroad tracks—the destination of the jacked tunnels.

The first completed jacking operation was on Ramp-D, an underground highway ramp. It was successfully jacked between June and December of 1999. It's the best example of the tunnel construction and jacking process.

Ramp-D is the smallest of the three tunnels jacked, but if it's the baby of the family, it's Baby Huey. The tunnel, 167 feet long, 78 feet wide, and 38 feet high, is like a 16-story building lying on its side. The Ramp-D tunnel is almost exactly the same size as the jacking pit in which it was built. With so little room, the contractors poured the concrete floor and walls of the tunnel, lowered construction equipment into the jacking pit, and parked the vehicles on the floor of the tunnel. With the machines needed for boring and digging now inside the structure, the contractors built the tunnel's roof over the top of the machines, encasing the gear until it was put to use months later.

When construction of the tunnel was complete, contractors spent another six weeks preparing it for its 180-foot journey under the railroad

of the family, it's Baby Huey.

tracks. They installed a powerful air-ventilation system, ramps, jacks, engines for thrusting the jacks, catwalks for the workers, and decking. They also set up thousands of feet of ADS rope (Anti-Drag-System rope) on large wooden spools along the roof and floor of the tunnel.

ADS rope is a 3/4-inch-thick lubricated metal cable. Without the ADS-rope, the ground above the tunnel and everything attached to it would be dragged along in the wake of the massive underground structure as it was pushed forward. The effects would be disastrous. Railroad tracks and high-tension poles for Amtrak's new electrical trains would snap and buckle. Trains would derail and underground utility lines would twist and snap. "It would be more disturbing than an earthquake," says a Local 88 tunnel worker.

To prevent this, contractors anchored the ends of the ADS-ropes to the jacking pit, with one group of ropes running on top of the tunnel and another under it. The tunnel was, in effect, riding on the ropes. As it was jacked out of the pit and under the rail tracks, the cables unrolled themselves from the spools, creating a friction-reducing layer between the tunnel and the ground. During the six-month process, the contractors periodically pump lubrication through the roof and floor of the tunnel. This reduces friction even further, limiting the damaging drag effect to the area between the tunnel and the cable. Instead of squeezing a greased pig through a picket fence, you have a 34,000,000-pound greased highway tunnel squeezing under the railroad tracks.

> *"The elevated highway is an eyesore. I was shocked when I moved here two years ago and saw how it cut the city off from the water and ruined the architectural beauty of some of the buildings. I'm glad I don't have to organize the funding for it, and I'm not well versed in the economic benefits or drawbacks of running the highway underground, but from an aesthetic perspective, it's a great idea."*
> —Greg Obenshain, Boston, MA

These men are working inside the 78-foot-wide, four-story-high Ramp-D tunnel. To the left, a member of Local 4 operates a roadheader, identical to one used in building the Chunnel under the English Channel, to clear a three-foot cavity in front of the tunnel before its next jacking. To the right, a sandhog watches a fellow worker in a man-lift remove a pipe used to carry liquid brine that is 30 degrees below zero and freezes the ground into a solid block of mud and ice.

Opposite:
Two sandhogs work late into the night hooking up a hose to a pipe 60 feet into the ground, with a massive refrigeration system which will freeze the ground into a solid block. This stabilizes the area and prevents the railroad tracks (background) from being dragged along with the huge tunnel section as it is jacked beneath busy commuter trains.

THE BIGGEST DEEP FREEZE

The biggest man-made freeze in history is taking place behind South Station. Beneath the train tracks and in front of the tunnels to be jacked, the ground has been turned into a frozen block. Calcium chloride solution, a liquid also called brine, is injected into pipes running 60 feet beneath the rail lines. The weak soil under the tracks would collapse during jacking operations if it were not first made into a solid block of ice.

In 1883 a German inventor patented the concept, known as soil freezing, to support coal-mining operations. No one, however, has used soil freezing on such a large scale before. Slattery, J. F. White, Interbeton, and Perini claim they save millions of dollars by using the technique over conventional grouting methods. It's another "largest" for the Big Dig. But more important, it works.

Refrigeration units large enough to cool the Empire State Building on a hot New York summer day worked for months to create a frozen soil block. "It takes three to four months for the ground to become solid," explains Rodwell, the engineer from London. "The brine travels through hundreds of pipes in the ground, back to the refrigeration units to be chilled, and then back into the ground. The brine remains liquid even when it gets to minus 30 degrees, and when it comes into contact with the ground it leaves it a chilly minus 15 degrees. It's quite remarkable."

Water expands upon freezing, pushing the railroad tracks up. Ideally the tracks move only five or six inches during the freezing process. Survey teams check track elevations twice a day, in the morning and evening, to ensure safe passage of the trains rolling a few feet from the moving highway tunnels below.

THE BIG PUSH

With the tunnel's construction inside the jacking pit complete and the outfitting of the tunnel finished, the behemoth was ready for its 180-foot-long journey. Roadheaders, identical to ones used in building the Chunnel between France and England, began grinding away at the frozen solid block in front of the tunnel.

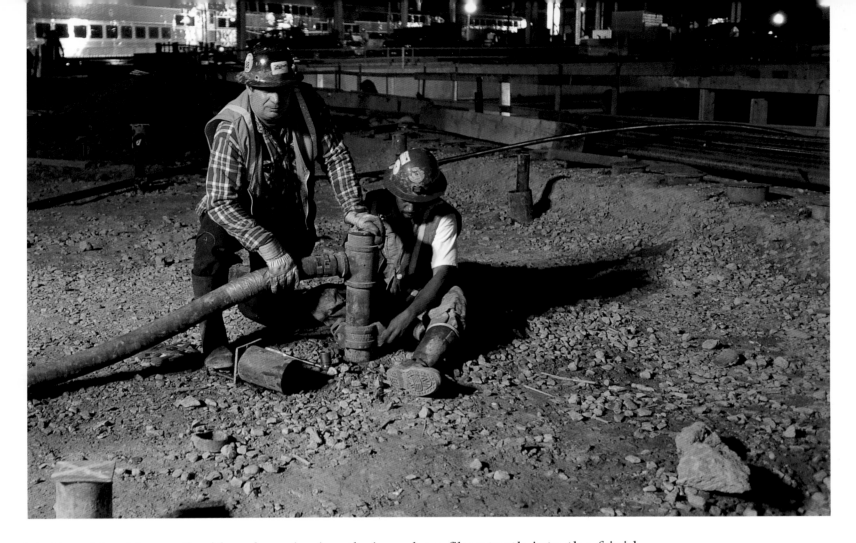

Medieval-looking spiked heads, spinning their carbon fiber teeth into the frigid material, tore away a three-foot cavity in front of the tunnel.

As frozen chunks of earth fell to the floor of the tunnel box, excavators swooped in and removed the dirt to a small pile. Special low-clearance front-end loaders then relayed that pile out the back of the tunnel and into the jacking pit. The still frozen load was placed into a large container, the size of a small Dumpster, and hauled up and out of the jacking pit by crane.

"It's a very dangerous work site in here because everything is happening so quickly in such a tight area. Inside these seven-foot-thick walls, it's about as big as a high school gymnasium, but instead of the girls' basketball team, you have these roadheaders, excavators, and guys with 5000-degree saline torches running around," shouted a Local 88 sandhog over the tunnel's blasting ventilation system.

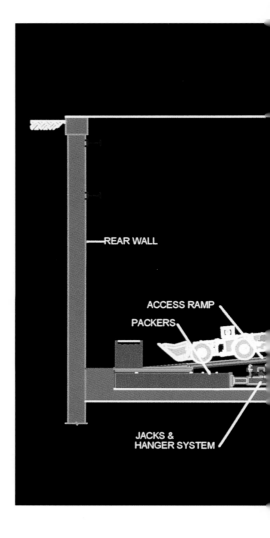

REAR WALL

ACCESS RAMP

PACKERS

JACKS &
HANGER SYSTEM

Opposite:
This diagram shows the Ramp-D tunnel being jacked under the rail lines behind South Station. The orange is the 169-foot-long tunnel, the purple is the wall of the jacking-pit, and the green train is leaving South Station. The two engines on the floor of the tunnel are the power packs, and the roadheaders are in the shield of the tunnel as it moves forward.

The race against time is not only about money. It's also about safety and productivity. If frozen material removed by the roadheaders is allowed to sit for fifteen minutes, it creates a slick mess on the tunnel floor, slowing activity to a crawl. Crews work quickly to remove the material in a precise underground ballet of movements. Otherwise work becomes hazardous, as melting heaps of earth create conditions comparable to a slick road in a winter storm.

With three feet of space chewed out between the Ramp-D tunnel and the frozen ground ahead, workers shut down the roadheaders and cleared the tunnel for safety reasons, leaving only essential personnel behind for the actual jacking. The crews fired up two enormous generators called power packs, seven feet tall and sixteen feet long, and engaged over 50 large hydraulic jacks, each capable of pushing 10,000 pounds per square inch (psi) of pressure. (Concrete walls break apart when just 4000 psi are applied to them.)

With the power packs roaring, the mighty hydraulic jacks attached to the tunnel pushed out against large steel tubes called packers, forcing 17,000 tons of concrete, steel, and equipment to slide three feet along greased ADS-ropes and under the railroad tracks. At the leading edge of the tunnels, nine-foot-deep shields penetrate the ground as the concrete box is pushed through the earth. Steel-tipped cutting bars attached to the shield's perimeters slice through whatever frozen ground remains, like an underground icebreaker cutting through the Arctic seas. Crews repeated this process over 60 times to get the Ramp-D tunnel to where it sits today, 25 feet beneath the trains arriving and leaving South Station every day.

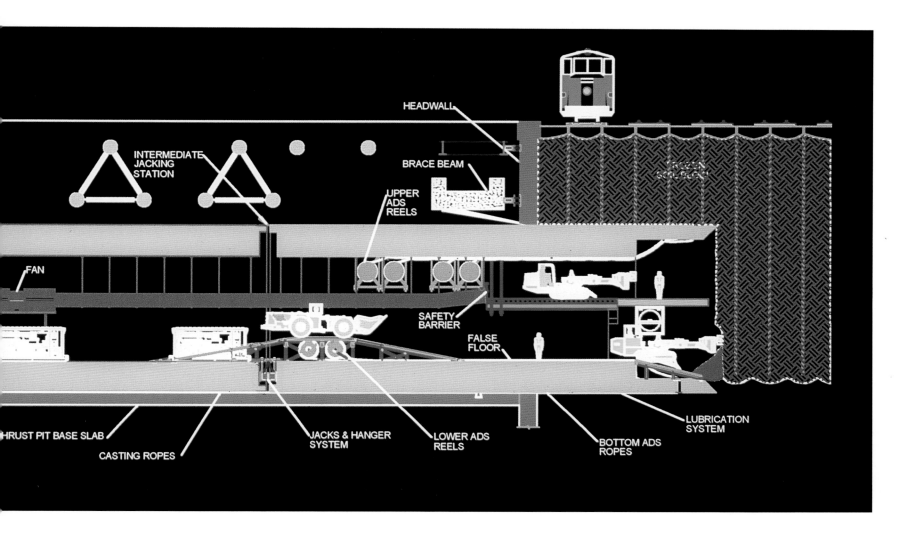

INTERMEDIATE JACKING STATION

FAN

HEADWALL

BRACE BEAM

UPPER ADS REELS

SAFETY BARRIER

FALSE FLOOR

THRUST PIT BASE SLAB

CASTING ROPES

JACKS & HANGER SYSTEM

LOWER ADS REELS

BOTTOM ADS ROPES

LUBRICATION SYSTEM

After pushing Ramp-D's tunnel to its final location, the contractor allowed the frozen ground to thaw. More than six months after the jacking and long after the brine stopped circulating, the ground around the tunnel was still frozen solid. Officials will keep a close watch on the thawing to ensure there's not too much settlement under the tracks.

After jacking 15,000,000 pounds of tunnel nearly 1000 feet, the Big Dig will have jacked more mass than has ever been jacked before. For commuters traveling to and from the airport, the work will represent a few seconds of travel time as they pass, unaware, through the tunnels.

CHAPTER TEN

The Big Dig's Biggest Challenge

T'S MISERABLE SOIL, PURE JUNK! It's a mass of weak clays and old landfill that's crammed with rubble and burnt timbers from the Great Fire a hundred and thirty years ago. It's possibly the worst type of material to build a highway tunnel through." So says Mike Bertoulin, area manager for the Fort Point Channel. But Bertoulin, whose grandfather was an engineer on the Panama Canal, doesn't shrink from his duties. "I love a challenge— and this is the project's biggest."

The Fort Point Channel crossing is certainly the most technically challenging piece of the Big Dig. The channel, once a little-noticed backwater, divides downtown from South Boston. The 1100-foot-long, 11-lane-wide tunnel through it is costing over $1,500,000,000, making it the most expensive highway per mile anywhere in the world.

Laboring through the night, Big Dig workers stabilize the weak soil of the Fort Point Channel by injecting grout into the ground. Wedged between South Station's commuter trains, Boston's primary U. S. Mail sorting facility and Gillette head-quarters, the Fort Point Channel operation requires precision construction work.

It has also been the source of much frustration for a time-pressured staff. "The channel crossing has become the catch in the drain for most of the Big Dig's problems," explained one engineer. "This is where we're playing catch-up. When time needs to be made up in the project's overall schedule, it's the nines [Fort Point Channel construction contracts] that are expected to do it." Never have 260 feet of shallow, muddy water been so difficult to cross.

In 1994 *The Boston Globe* ran splashy headlines claiming a $500,000,000 overrun on the Big Dig's Fort Point Channel crossing. Six years later, when the final numbers began to come in, it turned out they were close to the mark. Part of the largest, most extensive geotechnical investigation ever undertaken in North America revealed weak soils, weaker and far more unstable than earlier tests had indicated. Not surprisingly, questions arose about whether the Joint Venture had adequately tested the soil before things got this far. Just before the point of no return, the original plan was scrapped and $19,000,000 was spent on a new design. Criticism of the Joint Venture mounted, as newspapers had a field day.

Big Dig's acting project director, Mike Lewis, explains: "We were developing and preparing the final design, and we started to get information that in order to make a stable excavation, the costs were skyrocketing. We realized this was going to bust the budget of the project, so we said, 'Everyone out of the water.'"

Originally the Big Dig's plan to cross under the channel called for two concrete Immersed Tube Tunnel sections similar in concept to the Ted Williams Tunnel's steel ITT sections. On either end of the ITT sections, on the banks of the channel, were cut-and-cover tunnels. The land tunnels used cofferdams (support walls) to keep the Fort Point Channel's water out while workers built in the "dry." Keeping an area the size of Fenway Park's baseball field free of water during construction became astronomically expensive, especially when the newly discovered weak soils were factored in. Soft material is nearly impossible to hold back during construction. It is heavy and water-saturated, and stabilizing it involves costly and elaborate schemes.

The revised plan amounted to shrinking the portion of construction in the "dry" and increasing the work in the "wet." In other words, the total construction area to be dewatered was reduced from the size of Fenway's whole ballpark to the size of its infield. Since every action has a reaction, the work in the wet was increased from two concrete ITT sections with a total length of 300 feet, to six ITT sections with over 1000 feet in length.

There was one catch; some of the project's engineers were required to attend deep-soil-mixing class in Japan to learn how to turn soft ground into rock. After class, they embarked on the largest soil-stabilization effort in North America's history.

The Big Dig is a favorite target for satire.

Mike Lewis explains, in a 1998 interview, about the Big Dig and Japanese soil mixing: "In Asia, where there has always been a land crunch, using soil mixing, they have been able to reclaim very soft soils. They've been able to build an airport in Tokyo harbor on land that we would first think is prohibitive. We sent a team of engineers over to Japan to be with them and to learn their applications. We made adjustments, modified them to fit our particular needs and we're progressing very successfully."

In the fall of 1996, launching a $250,000,000 soil-stabilization campaign,

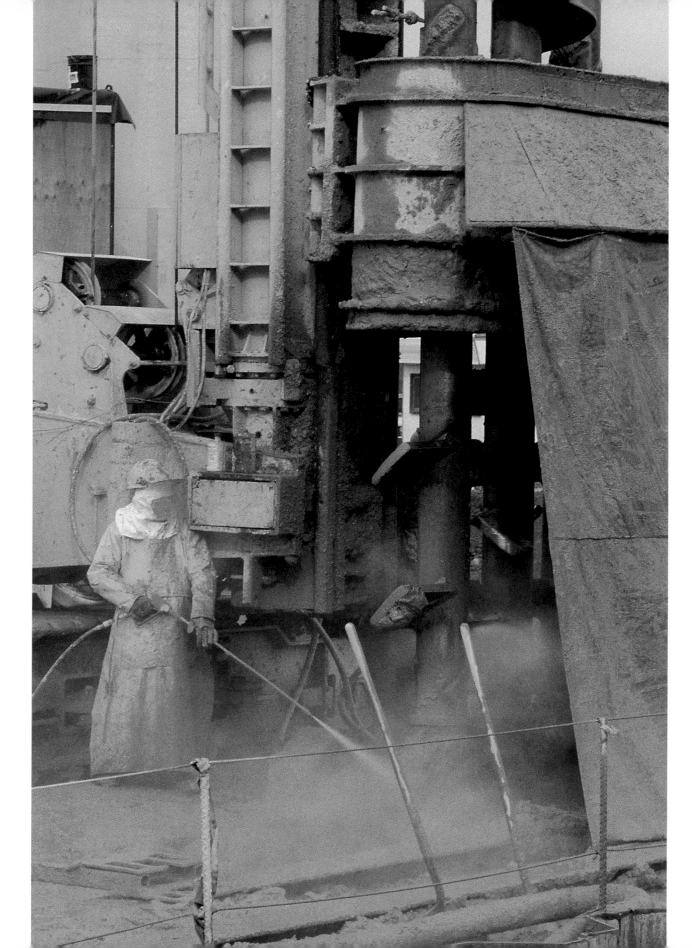

A deep soil-mixing machine works the weak soil of the Fort Point Channel. Big Dig officials went to Japan to learn how to turn unstable, soft soil into hard ground. A Big Dig worker washes off the "cake-paddle" mixer as it returns to the surface after taking a trip 120 feet underground. On its way down, the mixer injects water into the soil while the paddles churn the ground. On the return trip, the mixer injects grout into the watery ground, which hardens and stabilizes the area.

many of the largest deep-soil-mixing rigs on the planet began operations on the Fort Point Channel's western edge. Their 120-foot-tall booms became Boston landmarks as the machines worked the channel for over three years, each driving three overlapping augers that looked like giant cake mixers, deep into the soft soils. The Japanese-style mixing rigs churned the soft earth, penetrating 130 feet down to the first layer of glacial till, injecting water through the stems of their augers on the way down. The water liquefied the soil and prepared it for the next step, grouting. The trio of augers, upon reaching their destination deep into the ground, reversed gears and backed the rigs' giant churning paddles out of the hole, pumping cement grout into the watery mix as they went.

On the opposite side of the channel, a large grouting plant supplied the dinosaur-like rigs with all the portland grout they could pump into the soil. Silos delivered the grout through a network of pumps, pipes, and hoses, sending it to the roving machines, city blocks away. As the rigs backed the augers out of the ground, a solid column five feet thick, 15 feet wide, and 130 feet deep was left behind. The process was repeated thousands of times over many acres, transforming soft wet soil into a solid mass. The ground, filled with the light structural concrete, was now strong enough to support the excavation of one of the world's widest tunnels. Big Dig crews could begin to close the one-mile gap between the Massachusetts Turnpike and the Ted Williams Tunnel.

THE FORT POINT CHANNEL'S INDUSTRIAL ROOTS

The great industrial age of the United States lives on in the Fort Point Channel, a man-made, nineteenth-century waterway embodying Boston's maritime history. Seven old bridges and the remains of several more cross this historic body of water, making it a living museum of industrial designs from the last 100 years. A bascule bridge, a rim-bearing swing bridge, and a rare retractile drawbridge cross the channel within a few hundred feet of each other. Sitting in

The basin is another complication of the Big Dig, a project of

its water are old stone seawalls, timber pilings, wooden navigation fenders, and rusted pipelines.

In 1832, when South Bay was a saltwater basin and not a highway interchange, Cyrus Alger opened an ironworks foundry there, filling in part of the bay and starting the shaping of today's Fort Point Channel. In 1836 the Boston Wharf Company, with Cyrus Alger on its board, built docks and wharves in the channel to support the company's commercial shipping operations—importing and exporting sugar and molasses from Central America and the Caribbean. The Boston Wharf Company dominated the construction and management of warehouses and sugar refineries along the channel through the 1880s. Today, as the Big Dig breathes new life into the channel, Boston Wharf Company continues to manage over 70 of its original structures, leasing space to Boston's high-tech and financial firms.

The bulk of the original industrial activity was located at the southern end of the channel, near the South Bay Interchange and today's Big Dig construction. In addition to sugar refining, Fort Point was home to Boston's wool industry, heavy machine construction, coal storage, and distributors of lumber and other building supplies. In 1898, the South Station terminal opened and commandeered the channel's entire western edge, adding freight and passenger trains to the already frenetic marine traffic.

Today the United States Postal Service controls this side of the channel and South Station's presence has been reduced, offering passenger rail service only. In the heart of one of the smallest but most critical Big Dig contracts sits the post office's busy loading dock, where millions of pieces of mail are picked up and dropped off each day. Post office police stand guard over its two lanes of roadway, the lifeline of the facility, as hundreds of postal trucks roll through the severely restricted construction zone. Dump trucks, cranes, workers' pickup trucks, and mail trucks all maneuver for space on the tiny road.

seemingly endless complications

Tragically, in the spring of 2000, a Big Dig worker was struck and killed by one of these vehicles.

Across the channel, on the east side, Gillette's "World Shaving Headquarters" dominates the landscape. Here, Gillette, Boston's largest manufacturing employer, produces over a billion razor blades a year. Daily, hundreds of trucks make pickups and deliveries during the plant's 24-hour operation, while the Big Dig, also on a 24-hour schedule, constructs 11 lanes of super-highway beneath its property.

BUILDING A DRY DOCK

By the summer of 1997, David Sailors, a renowned photographer, had traveled around the world several times taking pictures for engineering and construction firms. Then a client asked him to come to Boston and photograph the Big Dig. Walking past Gillette's factory and up to the edge of a huge basin dug by Big Dig workers, his only words were, "My God, I have never seen a hole this big!" And he wasn't kidding. This canyon, 1000 feet long, 60 feet deep, and 300 feet wide, is big enough to hold the U.S. Navy's largest aircraft carrier, flight deck and all. It is another complication of the Big Dig, a project of seemingly endless complications.

Bethlehem Steel, the company that manufactured the steel ITT sections for the Ted Williams Tunnel and barged them up to Boston Harbor to be immersed, planned to do the same with ITT sections meant for Fort Point. But the narrowness of the channel eliminated that option. The widest ITT section of the six to cross the Fort Point Channel is 174 feet wide. If it were floated down the 260-foot-wide channel, there would be almost no room to spare. And that doesn't take into consideration shallow water, old bridges that no longer swing or lift, and other obstructions.

Building the ITT sections on-site, inside the Fort Point Channel, became the only practical solution; and for cost reasons, concrete was the material of choice. Americans are the world leaders when it comes to steel ITT sections. Concrete sections are a different matter. They have been built in Asia and Europe but never in the United States. This concrete ITT tunnel will be the first in the Western Hemisphere.

In the fall of 1995, work began on the construction of the basin where the tunnels, six in all, would be cast. The $158,000,000 contract required the contractor to build nearly a half mile of slurry walls 90 feet into the ground. Once the slurry walls were in place, thousands of dump truck loads of material were excavated from between the walls to create the basin.

ONE MILLION GALLONS OF RAZOR BLADES

In order to build the casting basin, Big Dig crews had to demolish and remove an abandoned storage tank with a capacity of 1,300,000 gallons of oil at Gillette's factory. "It was more like a bomb shelter than a fuel tank," explains Al Franzen, longtime field engineer in the channel. "It had thick concrete walls and was filled with reinforcement bars of every type and thickness you could imagine. It was nearly impossible to break up. Not only that, but it was filled with razor blades. I don't know how they got there, but the thing was filled with them, and we had to get rid of them all—oily razor blades. It was a mess!"

The abandoned fuel tank was only one of many obstructions Big Dig crews wrestled with as they prepared to dig the project's largest hole. Seven resin silos, a parking lot, fences, and Mount Washington Avenue—a local road-way—needed to be relocated or eliminated.

The casting basin is kept dry throughout the construction of the six ITT sections by a large dam. The dam is built, partially removed, and rebuilt to allow for the six ITT sections to be constructed and launched in two phases. The first phase was completed in January 2000 when the dam was removed and four completed ITT sections were sent out of the basin. In early 2001, the last two sections will be finished and the dam will come down one last time.

The only problem is the dam is so damn big. It is 600 feet long, as long as some World War II battleships. It is a series of six circular and seven kidney-shaped cofferdams. Sheet piles nearly 80 feet long are driven ten feet into the channel's mud, and thousands of tons of stones are poured into the cells or cofferdams to make a solid wall between the Fort Point Channel's water and the casting basin.

When the six ITT sections are resting at the bottom of the channel and the casting basin's dam is no longer needed, it will be completely removed. The original historical seawall, built 100 years ago, will be rebuilt exactly the way it was found when the Big Dig arrived. In the fall of 1996, under the watchful eyes of historians, crews removed each granite block, carefully painting identification numbers on each stone so it could be put in exactly the same location six years later. Currently, the old seawall sits in the Rumney Marsh, a saltwater wetland ten miles north of Boston, awaiting its return.

In the summer of 1997, after 27 months of work, the casting basin was ready for production. Before the last of the dump truck loads of dirt

Work on the opposite side of the channel from the casting basin held up the launch of the ITT sections. Working around the clock, this clamshell bucket dredges the bottom of the channel where the first two of the six ITT sections will come to rest. The two silos in the back right store the grout to be injected into the ground.

181

were removed from the basin, the contractor began building the first concrete ITT sections in America—which are also the world's largest.

BUILDING A DRY DOCK

The 1000-foot-long casting basin, with its 600-foot-long dam and half mile of slurry walls, quickly morphed from Grand Canyon to dry dock. To make the fabrication process a waterless one, the floor of the basin is equipped with its own drainage system, pumping out 100 gallons of seawater and groundwater per minute. Six inches of crushed rock is placed over the bottom, allowing water to seep under the completed superstructures when the basin is flooded, thus preventing the cohesion that would cause the ITT sections to stick to the bottom of the basin. Three inches of sand is placed over the stone to create a perfectly level surface for casting the concrete sections. Precision is a constant theme throughout the two years of construction of the giant ITTs.

Every item going into the ITT sections is weighed, recorded, and calculated in preparation for the day the section will be floated out into the channel and immersed. Every ounce of concrete, plumbing, electrical work, and even minor pieces of material are factored. If the walls are too thick, the ITT won't float; if they are too thin, it won't sink. Knowing the exact weight is crucial to predicting the ITT's behavior.

On top of the sand, large steel plates are welded together, forming a protective shield for the bottoms of the ITT sections. Without the shields, objects could puncture the undersides during float out (when the ITTs are floated into the channel and immersed).

The ITT sections' construction is similar to the tunnel boxes being built underground in downtown Boston. The floors, walls, and ceilings are cast in place as millions of cubic yards of concrete are poured into forms that give the tunnel its rectangular shape. Each pour is allowed to cure while the work

Opposite: Digging out the massive casting basin next to Gillette's headquarters in the summer of 1997. A backhoe fills a convoy of dump trucks with earth. The casting basin's dam holds back Fort Point Channel's water and provides a convenient work site for the large yellow crane operating a clamshell bucket.

Opposite:
Gillette's pumps cool water from the Fort Point Channel into its facility through large utility pipes that cross over the casting basin. An engineer walks over the basin on a catwalk while another worker inside the casting basin operates from the top of an ITT section under construction.

of building the highway tunnel sections continues.

Carpenters, operating out of their own mill at the bottom of the basin, provide ladders and railings for safety and build forms to support newly poured concrete. Crane operators take apart 200-ton cranes and lower them six stories into the bottom of the basin and rebuild them, so they can support the other trades by moving materials and gear around the casting basin. Topside, their colleagues support the operation with even larger cranes operating along the basin's edge.

Over the three-and-a-half-years of work, thousands of men and women will have performed tasks not often associated with tunnel building. Laborers, cement masons, pipe fitters, sheet-metal workers, plumbers, electricians, and sprinkler fitters gave critical support to the assembly of these unique, world-class behemoths. "It takes a village to build these things," joked one tradesman.

Final preparations for the flooding of the basin and the first float out began in the fall of 1999. Large wooden stalls, one in each corner of the tunnel section, were built and lined with rubber, so thousands of gallons of water could be pumped into them. Bigger and deeper than your neighbor's swimming pool, these ballast tanks provided engineers with a mechanism for leveling the sections once they became buoyant. Once the ITTs are floated to the surface, water in the wooden ballast tanks can be pumped in or out to trim the floating sections. If a section is listing to its right side, water can be pumped into a tank on the left, keeping the gargantuan section floating evenly.

Hulking steel bulkheads attached to the ends of the completed sections keep them airtight and dry inside. Two layers of water protectant were sprayed over the four sections. Each layer has a different color, yellow or gray, so any missed spots in the spraying can be addressed. Large rubber gaskets placed on the ends of the sections ensure a successful vacuumlike connection when the six ITT sections are finally brought together under the murky Fort Point Channel water.

PREPARING THE CHANNEL—PREVENTING A FLOOD

In order to place the massive sections on the bottom of the channel, nearly as much material needed to be cleared from the bottom of the channel as was excavated from the casting basin. While work inside the casting basin progressed on the ITTs, dredging operations were under way in the channel, removing 350,000 cubic yards of material from the muddy water.

Removing that much slop from water with zero visibility is difficult, but when Boston's second busiest subway line passes within five feet of your bucket, you need to be especially careful. The Big Dig is digging, drilling, and laying nearly 100,000 tons of highway tunnel only six feet from the roof of the Red Line subway.

"There are more instruments in the Red Line tunnel than in a 747 jumbo jet," proclaims John Bales, who moved to Boston from Britain to oversee the construction of the ITTs. "We weren't even excavating near the thing when we got a phone call from Seattle, where they monitor this equipment, asking us what was going on with the subway tunnel. Their instrumentation picked up a slight movement from a dewatering operation over a thousand feet from the subway tunnel. Nonetheless, the movement of groundwater triggered their gauges!"

Meanwhile, back in the channel, lasers from the shoreline guide the dredging bucket 22 feet down, preventing an inadvertent hit of the subway tunnel. John Bales describes the scene: "We shoot a laser light from the embankment out to the dredging bucket so the operator always has a read on the distance between the roof of the subway tunnel and his scoop. It takes the guesswork out of the equation because the laser beam is set at a constant elevation, and that eliminates the need to factor in the rising and falling tides. It's just another insurance policy against Damageville!"

Near the dredging bucket, work barges support construction crews

Opposite:
In the summer of 1997 the casting basin begins to take shape. Looking west, toward the Fort Point Channel, work crews continue to dig out the basin while crushed rocks are laid on the basin's floor under pipes used by Gillette to cool manufacturing equipment.

drilling 110 shafts deep into the Fort Point Channel's bedrock. These drilled shafts are six feet wide and extend down 190 feet from the surface of the water. Penetrating 18 feet into bedrock, the shafts will act as underwater columns, holding up the first four ITT sections.

It's another series of firsts in the world. ITT tunnels have never been fully supported by drilled shafts, never before have they been placed so close together (five feet apart in some cases) and never has a subway train traveled through a field of drilled shafts.

FLOOD THE CASTING BASIN—FLOAT THE TUNNELS

More than two years after construction began on the first four ITT sections, they were ready for launching. The basin was dry, the sections were waterproofed, and their open ends had been sealed off with large steel bulkheads. Few people realize, however, that the foundation of a colossal ventilation building and bridge piers for a four-lane road were attached to the top of the first two sections to be floated out of the basin.

The ventilation building's 22-foot-high basement walls were attached to the front portions of two of the ITT sections. Also on top of the sections and beside the vent building's wall, were the 22-foot-tall columns that would hold up the new Dorchester Avenue Bridge, to be built by the Big Dig and used by the United States Postal Service.

"Only Americans would be crazy enough to put a building's basement on top of an ITT!" joked Bales. Crazy or not, it has a purpose. A tight construction schedule and cramped workspace near the postal facility on the western side of the channel brought on the challenge. Building the supports for the basement walls and bridges in a bone-dry environment reduced underwater construction and helped keep the schedule on its critical path. However, it put tremendous additional pressure on the crews who navigated the ITT sections across the channel.

Opposite:
In 1998, about a year after the previous picture was taken, the casting basin holds four of the six ITT sections that will be built inside its walls. All four of these sections have been floated out of the basin.

Meet "Big Jim" Harris, bargemaster extraordinaire, recruited to Boston for his unique talents in barge handling. His on-the-job experience in navigating unwieldy barges down uncooperative rivers qualified him for this difficult task. His job of delivering nearly 100,000 tons of concrete and steel meant the entire Big Dig schedule was in Big Jim's hands. If the sections were damaged during their trip across the channel, the project's entire time frame could have been derailed. Jim literally rode the tunnels across, radioing commands to tugboat operators and linesmen on shore.

Before he could get started, the casting basin needed to be flooded and two thirds of the dam taken apart to allow the four ITT sections enough room to squeeze out. Before the basin's flooding, wooden ballast tanks inside the ITT sections were filled with water, acting as heavy weights. Tall cylinders, called stability tanks, were temporarily placed on top of the sections and, like the tanks inside the sections, they were filled with water to provide additional ballast, ensuring that the tunnels would come to the surface only when engineers wanted them to. Then water was pumped over the cofferdams and into the basin. After three days of pumping water, the casting basin was filled and the ITT sections were under three feet of water.

Work across the channel progressed slowly—in fact, too slowly. It was holding up the launching of the tunnel sections. The delay was due to the precision the job required. Before the sections could be brought across the channel and lowered into place, the drilled shafts needed to be perfectly level, so the ITTs would fit evenly across their tops. With hot water pumping through their diving suits, divers worked 24 hours a day with high-powered water guns blasting away at uneven concrete around the tops of the drilled shafts.

With near zero visibility, the divers were in constant audio communication with a dive barge, giving each other critical details about location and condition of the shafts in the water. "Sometimes you turn this light on [the light attached to dive helmets] and it makes it more difficult to see. If I can't see anything, I'll just close my eyes while I'm doing my work. It helps, because I'd rather depend on my sense of touch than sight down there," explained one diver.

In January 2000, 41 drilled shafts were level and ready for the first ITT section to be placed on top of them. "It's a little nerve-racking preparing these shafts on one side of the channel to fit perfectly into the bottoms of the ITT sections on the other side. We only have one chance at making the 41 connections fit, and we don't find out how accurate we are until the operation is nearly over," explained Bales.

PREPARE TO LAUNCH!

Thousands of gallons of water were released from the interior wooden ballast tanks to lighten the load of the 46,000-ton goliath still resting at the bottom of the basin, still underwater. Bright blue polypropylene lines connected the ITT section with large diesel-driven winches on the sides of the casting basin, preventing the section from drifting and colliding with the other ITT sections once it reached the surface. Finally, the water trapped in the yellow stability tanks on the top of the ITT section was pumped out and the section slowly rose to the surface. Once the section roof was four inches above the surface, engineers adjusted the water in the interior wooden ballast tanks to reach the desired pitch and yaw. With the section level, crews waited for just the right day to launch.

On Sunday, January 9, "There wasn't a breeze. The water was dead calm, like glass. The conditions were ideal for a launch," recalls Mike Bertoulin. The polypropylene ropes now extended across the channel in order to pull the ITT section to its final destination.

Bargemaster Big Jim was in the control house on top of the ITT section,

Overleaf:
The basin is over 1000 feet long, 300 feet wide and 60 feet deep, large enough to dry dock the USS John F. Kennedy, one of the Navy's massive aircraft carriers. In the spring of 2000 the basin is empty and about to be drained so the last two ITT sections can be built.

In 2002, I-90 will finally connect Seattle, Washington, to Logan

armed with enough instrumentation to rival NASA's mission control. The call came at 9:30 p.m. In the pitch-dark, the diesel engines driving the winches started pulling their lines and the section began to move out of the casting basin, across the channel, reaching a top speed of 20 feet per minute. To ensure a safe passage over the Red Line, the trip was timed to float the section above the roof of the subway at high tide. After clearing the subway's tunnel, Big Jim began to concentrate on the hard part—delivering his 46,000-ton baby to its exact location, connecting four, three-foot-wide pin-jacks in the extreme corners of the ITT section with four drilled shafts in the dark water below.

The ITT section and Big Jim were in constant communication. The floating fortress of concrete and steel was rigged with antennas, transmitters, and eight Global Positioning Systems. The GPS units bounced information off a constellation of 21 orbiting satellites, of which four were constantly reading information to Big Jim. The software on board the ITT was able to interpret the GPS data and project a real-time display with graphics and text for Big Jim to follow. Pictures of his winch lines, speed, and target position were updated once every second with astonishing accuracy—to within one third of an inch horizontally and three quarters of an inch vertically.

Nine and a half hours later, just before sunrise, the ITT section was in position over the drilled shafts. The lowering tide slowly brought the enormous structure closer to the shafts on which it would rest. Big Jim deployed the four pin-jacks inside the ITT section. Slowly, pistons extended out of the bottom of the tunnel section and projected themselves into the tops of the shafts. At this point, more water was pumped into the internal wooden ballast tanks to add weight, guaranteeing that a rising tide would not lift the section off the shafts it had just come to rest on. "We had some luck. . . . I think we were within three eighths of an inch [of perfection] all the way around," Big Jim told *The Boston Globe.* At precisely 6:54 Monday morning the section touched down.

International Airport...

From inside the section, grout was injected into the four outside corners, ensuring a solid airtight connection. Divers on the outside inspected the injected grout, making sure it set properly, before the rest of the shafts were grouted to the ITT. Then Mike Bertoulin celebrated with his crews.

THE MISSING LINK

In the spring of 2000, the casting basin was re-dammed and drained, and construction on the last two ITT sections began. The first two sections were immersed in their final resting spots and the other two are moored next to the postal service complex. They will remain tied up and floating until they are placed behind the sections now being built, sometime in 2001.

Just beyond the casting basin's wall and before the Ted Williams Tunnel, another $500,000,000 of construction is underway. In 2002, I-90 will finally connect Seattle, Washington, to Logan International Airport, completing the original interstate highway plan started so long ago.

Overleaf:
The missing link between the Fort Point Channel and the Ted Williams Tunnel (left center) is a $500,000,000 construction project. An 11-lane superhighway tunnel will complete the original plans of the United States Interstate System, by extending it from Seattle, Washington, to Logan International Airport. At bottom center, the Big Dig is building the new Silver Line transitway/subway tunnel and station.

CHAPTER ELEVEN

The Charles River Crossings

MERGING FROM THE GROUND, soaring to 330 feet above the water is Boston's future. The old industrial seaport of Boston has come back as a world-class city, rich in history, culture, business, and finance. The Charles River cable-stayed bridge, a towering angular mass of steel, concrete, and cable, is the trophy that proves it. It pierces the skyline of the city, jutting out among the red bricks and granite buildings that define this old port town—a bridge that has taken the lead in the highly visible world of bridge design.

It is the crown jewel of the Big Dig. It was fought over, kicked about, argued, and finally resolved in a classic theater of debate, anger, and pride, Boston-style. With a sister bridge nearby, the Storrow Drive Connector, it will provide a total of 14

The Big Dig's cable-stayed bridge, still under construction, has already become a Boston landmark. White cable-stays reach up to the bridge's south tower directly from one of its two back spans. A metal staircase allows workers to climb to the top of the bridge tower.

lanes of traffic across the river, making logical connections and replacing six cramped and aimless lanes of an antique trestle bridge.

The name River Charles first appeared in 1614 on a map of Boston drawn by Captain John Smith. Smith wanted the political goodwill of the 15-year-old English prince who would soon be King Charles I. Naming a river after the heir apparent seemed an easy way for the New World to stay in the good graces of the old one. Born of politics, the Charles River has never been a simple river to cross. For hundreds of years, great and not-so-great Americans have struggled for the best way to navigate it.

In 1786, ending 66 years of debate, John Hancock and 83 private investors built Boston's first bridge, the Charles River Bridge, from Boston to Charlestown. It was 1503 feet long and 42 feet wide. Seventy-five piers of solid oak timber supported it as it spanned the rushing waters of the Charles. There was a convenient six-foot passage for pedestrian traffic on each side and at night the glow of 40 lamps provided illumination. It cost 15,000 pounds, an enormous sum at the time.

In 1997, more than 200 years after John Hancock's bridge opened, construction began on the new Charles River Bridge. It is the result of thousands of community meetings, 56 designs and redesigns, billions of dollars, and years of political battles and court actions.

SCHEME Z

In 1988 engineers working for Secretary of Transportation Fred Salvucci were in a fierce debate about how to cross the Charles River. There were 31 different schemes, going under the river, over it, or both. Scheme S designs were river crossings with both tunnels and bridges, Scheme T consisted of just tunnels, and Scheme Z proposed tunnels and bridges. The engineers argued heatedly about the merits of their designs and choices. Meanwhile, an unfriendly Republican White House was applying pressure on Salvucci's team to get the Big Dig under way or lose its funding.

Salvucci found himself in the middle of an argument that threatened to pull the plug on the Big Dig. Without a consensus, he made the decision himself and chose the least expensive plan: an all-bridge crossing called Scheme Z modified, or just Scheme Z.

Even though Scheme Z worked from an engineering standpoint, it was not a popular choice. It was the most environmentally sound, the least disruptive to the polluted river bottom. But it proposed more ramps than any other design and reached 110 feet into the air, with 18 travel lanes on six levels crisscrossing the Charles River. Pulitzer Prize-winning architectural critic Robert Campbell later called it "an awful scheme for a great wall across the Charles." Steve Ells from the EPA said it would be the "single ugliest structure in New England." Even Salvucci's staff members at the Central Artery/Tunnel Project resisted it.

In early 1991, with Salvucci out of office, Tom Larson of the Federal Highway Administration continued to pressure Massachusetts to find a solution to Scheme Z's overly complex design. In early 1991 it was scrapped and a new plan was in the works. The communities participated through a newly formed 42-member Bridge Design Review Committee with representatives from the neighborhoods, business community, and environmental groups. The large committee was considering at least nine bridge designs, with a great deal of mistrust and suspicion among members over the merits of each design. Former Big Dig manager Stan Durlacher remembers that at first the state was clearly the villain. "We could have had our engineer stand up and say 'The sky is blue,' and we would have a tremendous debate about 'How blue is blue? We don't believe it's blue. How do we know it's blue?'"

By late 1993, the committee had decided to support an expensive plan called 8.1.d Mod 5. It was part bridge and part tunnel and would potentially cost $800,000,000 more than Scheme Z.

The final word, however, belonged to the commonwealth's Secretary of Transportation, James Kerasiotes. And he was quietly reviewing designs with his own team of engineers. They were looking at no less than 16 bridge plans—arches, trusses, suspension, and six types of cable-stayed. Explains Vijay Chandra, one of the engineers leading the Big Dig's selection process, "We weighed the options very carefully and broke out their costs to a per-square-foot number. Most important to us were aesthetics, functionality, and cost. We then quickly reduced the field to seven bridges: a simple span bridge, a two-hinged arch bridge, and five cable-stayed designs."

The seven contestants proceeded to the next rigorous round of testing—site selection. With a location full of challenges, the seven contenders dropped like flies until only two were standing. These final two contestants were both able to accommodate a subway tunnel traveling through their foundations, a road surface with a severe pitch, the need to build around the old I-93 bridge, and heavy boat traffic throughout construction. But one was deemed better than the other, and it is the bridge now being built.

Kerasiotes asserts, "Forget about 8.1.d Mod 5. Since we don't want Scheme Z, we found some alternatives that were in between. Any reasonable, rational review would ask 'Did we improve on Scheme Z?' Absolutely. 'Did we create more parkland?' Absolutely. 'Did we lessen the visual impact?' Absolutely. We did all the things we needed to do."

The decision was final, and to make a point about how expensive further delays would be, Kerasiotes reported another major increase in the Big Dig budget the same day he declared the Scheme Z issue resolved. The total budget had risen from $6,400,000,000 to $7,700,000,000. The city of Cambridge sued for another redesign, but the courts said "enough." In 1997 the Big Dig began work on the long-awaited river crossing.

THE WORLD'S WIDEST CABLE-STAYED BRIDGE

It is the widest cable-stayed bridge in the world, carrying ten lanes of traffic. Its main span is 745 feet. It is also the first asymmetrical (two additional lanes cantilevered onto the east side) cable-stayed bridge in the world, and the first hybrid (steel and concrete) cable-stayed bridge in the United States. If that isn't enough, it's a beautiful bridge.

To better understand what a cable-stayed bridge is, it helps to realize what it isn't—a suspension bridge. The Golden Gate Bridge is America's most famous suspension bridge. Suspension bridges are best suited for crossing large expanses in a single span. Two usually enormous towers, with two long and thick suspension cables draped over the tops of each tower, form a suspension bridge.

Cable-stayed bridges eliminate the large suspension cables. They usually have one or two towers, from which cables are attached to the bridge deck. The decks are usually made of concrete with steel cables. The Charles River Bridge is unusual in that its spans are not uniform; there is a composite steel main span and post-tensioned concrete back spans. This makes it the first hybrid cable-stayed bridge in the United States.

It has two towers—the north tower on the Cambridge and Charlestown side of the river and the south tower on the downtown Boston side. Each tower rests on 100-foot-deep foundations called drilled shafts, similar to the ones in the Fort Point Channel but larger by two feet.

The man behind the design of the crossing is Christian Menn, a renowned Swiss engineer. Menn is internationally known for his innovative cable-stayed and box-girder bridge designs. Working with other bridge consult-

Reinforcing steel used in the project would make a one-inch steel bar long enough to wrap around the earth at the equator.

ants, he took the lead position in making the bridge crossing work.

When designing his superstructure, Menn became enamored of the historical significance of the bridge's location. It's adjacent to the site of Paul Revere's supposed landing, the country's oldest navy yard, and the home of the tall ship known as Old Ironsides (U.S.S. *Constitution*). In Charlestown, less than a mile away, the Bunker Hill Monument marks the battleground of the famous Revolutionary War battle. Because he was so impressed with the monument's symbolism, Menn added two subtle obelisks, the shape of the Bunker Hill Monument, one at the top of each bridge tower.

In addition to being beautiful and in tune with its historic surroundings, the Charles River Crossing bridge comes with a reasonable price tag—reasonable, at least, for the Big Dig. The 1457-foot-long bridge will cost just over $100,000,000, considerably less than the tunnel project down the street.

At the extreme left of this photo, taken in the winter of 1998, the upper and lower decks of I-93 carry traffic over the Charles River. To the right, beneath the tower crane, the south tower of the new Charles River Crossing takes shape. In the center, the red crane places huge tub girders of the Storrow Drive Connector Bridge over the Amtrak tracks at North Station.

THE BABY BRIDGE

Even as all the planning and controversy was whirling around the cable-stayed bridge, another Charles River crossing was being constructed. Since the 1950s, state officials had been trying to build a bridge connecting I-93 to Storrow Drive, a local four-lane parkway. Affluent Bostonians in nearby Beacon Hill and Back Bay resisted, believing too much access between these busy roads would choke their neighborhood streets with traffic. By 1994, however, they were eager for an improved system.

In April of 1997, Big Dig engineers were surprised to see a steel box-girder bridge on the list of contenders for the four-lane Storrow Drive Connector Bridge. They were even more surprised when it was selected as the low-bid contract. "Concrete is usually much cheaper to build with and we didn't think a steel bridge had a chance of winning the bid," explained a Big Dig engineer. "I, for one, am glad; it's a fantastic structure that blends in with its [cable-stayed] sister. The steel gives it a great look. If this thing were sitting by its lonesome it would be a standout. As it is, it complements the bigger one."

The Storrow Drive Connector Bridge, as it is officially called, was often referred to as the Baby Bridge during its planning and early stages of construction. Upon delivery of its 2100 tons of giant tub girders, the nickname "baby" was rarely heard. Tub girders, or box, girders are hollow steel structures that rest on top of bridge piers planted deep into the river and its embankment. They are light and strong, with a concrete road deck—carrying four lanes of traffic—resting on top of them.

The nine jumbo tub girders were fabricated in Florida by the Tampa Steel Erecting Company. In the summer of 1998, they were placed on barges and sent through the Gulf of Mexico, the Florida Keys, and up the east coast to Boston. Meticulous planning in the design stages allowed the barges and tub girders to fit narrowly through thin tidal locks. With only inches to spare, the

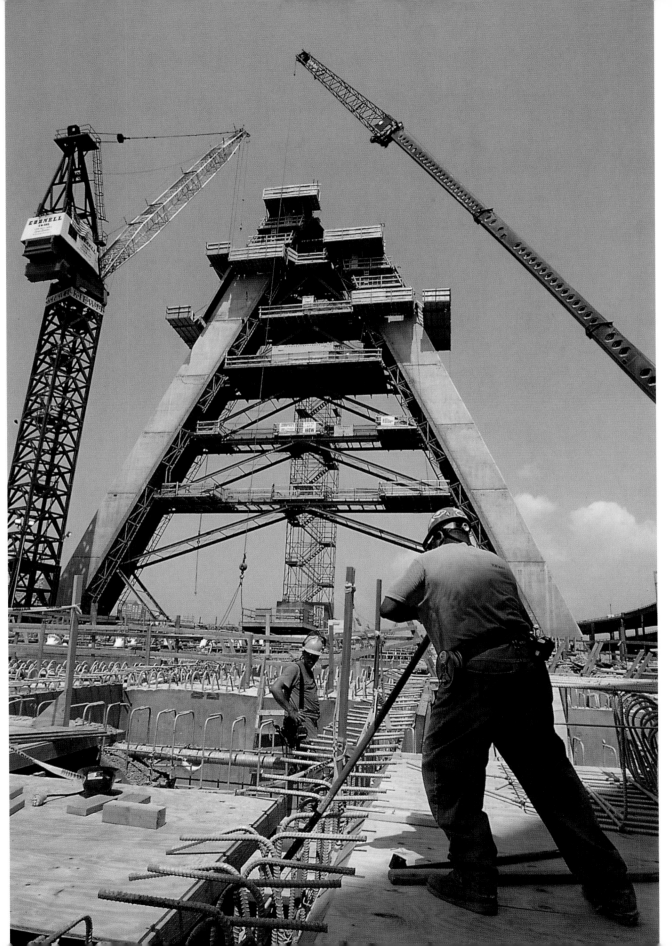

The south tower of the cable-stayed bridge begins to take shape in the spring of 1999. The "free standing" tower crane to the left has four foundations. Two of them are anchored into the ground; the other two are attached to the foundation of the bridge's tower.

207

Previous page: Big Dig staff often called the Storrow Drive Connector Bridge the Baby Bridge, always in the shadow of her big sister, the cable-stayed bridge. Workers in the summer of 1999 labored 24 hours a day, seven days a week to meet the October 1999 opening date. A large yellow gantry crane from Norway builds the back span of the connector bridge. The construction of the cable-stayed bridge can be seen in the lower right corner.

big tubs squeaked through the dam's locks and into the mouth of the Charles River, the bridge's new home.

Kerasiotes, then chairman of the Massachusetts Turnpike Authority, pushed contractors to their limit to finish the bridge. On October 7, 1999, weeks ahead of schedule, it opened to the public.

Big Dig traffic engineer Glen Berkowitz told *Banker & Tradesman* that he remembers the first few cars driving slowly down the ramp. "Drivers rolled down their windows and said, 'Thank you, thank you.' Usually when they roll down the window, they have less pleasant things to say." Berkowitz should know. He and his team have been hearing from drivers as they've been narrowing, shifting, closing, and reopening roads around the Big Dig for nearly ten years. The commuters wasted no time adjusting to new patterns. "I never underestimate the intelligence of Boston commuters," he says. "As soon as it opened, people started using it as if it had been there forever."

THEN AND NOW

The existing bridge over the Charles River is scheduled for demolition before the Big Dig is finished in 2005. Traffic reports call it the upper and lower decks of I-93, but its real name is the High Bridge. It won't be missed, no one has handcuffed himself to it in protest of its removal, and SAVE THE BRIDGE signs are noticeably absent. The High Bridge has been an unimpressive structure from day one and, some believe, poorly built. It is an ugly 50-year-old dilapidated structure that is literally falling apart. At about 3:00 p.m. on Friday of Memorial Day weekend in 1999, the upper deck of the bridge actually split from the roadway by about eight inches. The state police received reports that motorists were experiencing blowouts on the interstate and quickly traced the problem to sharp finger joints that had become exposed on the bridge deck's surface. Three of four rusted stringer beams that supported the upper deck had failed,

and within an hour nearly every street in downtown Boston was gridlocked. It was a stunning example of the damage caused by the overuse (400 percent of capacity) of an outdated design.

The new bridge is meant to last 100 years with proper care, as is the rest of the Big Dig's infrastructure. To show it off but not wash it out, the bridge is lit with 75 percent less lighting than other signature bridges. The lighting experts didn't want to create a spectacle, flooding the sky and the

When the carpenters, ironworkers, and laborers are finished building the south tower (above) they will pack up and cross the Charles River to finish the north tower.

211

neighborhoods with artificial light. White lights will shine up the cable stays and indirect blue lights, already installed, splash the towers.

On the deck of the bridge, large diamond-shaped openings have been cut out by Christian Menn and Company to allow daylight for boaters, bikers, bladers, pedestrians, and fish. Environmentalists feared that migrating fish starting up the Charles River from the harbor would become disoriented under the bridge's shadow and turn back out to the ocean.

A $96,000,000 park is being built underneath and around both bridges. Seven miles of bike paths and walking trails through more than 40 acres of park space will finally connect the Charles River with Boston Harbor, fulfilling the dreams of many citizens and urban planners. And the bridges will fulfill the fantasies of frustrated commuters—replacing six narrow, dangerous lanes with 14 wide and safe ones.

Computer image of the Charles River Crossing. In 2002 the cable-stayed bridge will be open to traffic. In 2005 the upper and lower decks of the old I-93 bridge (not seen here) will be demolished and the network of parks beneath both bridges will be complete.

CHAPTER TWELVE

The Home Stretch

HE MAGNITUDE AND SIGNIFICANCE of the Big Dig project appears to be understood best by the men and women building it. Every day thousands of people working in the trades report to their jobs on the Big Dig. Their work is some of the most dangerous in the country, as they scale walls of steel and concrete, descend into cold dark water and 100-foot-deep shafts, always risking injury or death. They seem to enjoy it, taking pride in their place in history. Many of them plan on telling their grandchildren about being a part of the largest construction project in the United States.

A worker gives a hand signal to a crane operator who is about to lower another load of reinforcing bars. Concrete will be poured around the reinforcement bar to make a solid highway tunnel wall.

"It all comes down to this, an ironworker tying together rebar with wire and his own hands," John Nordell, a photographer for *The Christian Science Monitor*, remarked, "I find it astonishing, all this

money, planning, and equipment, and none of it can get done without the actions of one person."

The Big Dig absorbs the labor of as many as 4500 trade workers and contractors during its peak years of construction, 1999–2001. Engineers, carpenters, bricklayers, ironworkers, laborers, and tunnel workers are only some of the trades that climb out of bed and go off into the rain, snow, and heat to risk their lives and build a piece of roadway. On any given night there are 1500 Big Dig workers pulling a night shift or working through a tough 12-hour day. "For four years I have been seeing my four children for two hours a day, between 3:30 and 5:30 in the afternoon," admitted Jim Marinellia, a tunnel worker 120 feet under Atlantic Avenue. "I just said good night to my ten-year-old daughter, Elena, on my cell phone. It's hard." Their dedication and endurance personifies what this project is all about—the people.

When the Big Dig was just a hope and a prayer, Boston's trade unions struck a deal with the commonwealth. They agreed to avoid strikes if the Dig hired strictly union workers. An agreement was reached, and the deal has been honored since its inception. The result—contractors have a highly skilled and competent workforce to draw from, making the work sites safer and more productive. The union members have enjoyed years of uninterrupted employment and experience. And there have been no strikes. With more work to go around than workers, times are good and people are breaking into new areas of expertise, becoming proficient with techniques and gear that wouldn't have been available to them unless there was a Big Dig.

CONTRACTORS AND CONTRACTS

Over 400 design and construction firms are working on the Big Dig through the peak years of building. Some of them are among the largest construction firms in the world, and others are small local companies with a once-in-a-lifetime

Opposite: With stickers of Stormin' Norman Schwarzkopf and the flag on his hard hat, is there any doubt this worker is a proud American? Hard hat stickers are like cigarettes in WWII. They are a type of currency between workers and are traded with a great deal of fanfare. A worker's hard-hat's color often indicates the trade he works in.

Opposite:
Every day, 4,000
workers apply
their skills on the
Big Dig, completing
$3,000,000 worth of
work every 24 hours.
The Big Dig has set
a new standard in
workers' safety, with
a very low fatality
rate. Three deaths
have occurred during
the first ten years of
dangerous work.
In three years 20 lives
were lost while
building the Golden
Gate Bridge. During
the construction of
the Panama Canal,
over 5,000 lives
were lost.

chance to work on an international project. Many of the smaller firms have received special assistance from the Big Dig because they are minority owned and operated, giving women and men in these firms a chance to partner with larger organizations from around the country and the world.

With a project of this size there is a great deal of risk . . . and reward. To minimize one and increase the other, firms often form partnerships called joint ventures. Joint ventures increase the contractor's chance of making a profit and lessen the likelihood of losing everything. Companies that often compete against one another for work on other projects have found themselves together in joint ventures on the Big Dig.

Even to bid for a Big Dig contract, a firm must first prove it is qualified, producing the necessary credentials to the officials of the commonwealth. For example, if a contractor wanted to bid on a design contract, he would need to produce his engineering or architectural licenses, whichever applies to the potential work, in order to be considered qualified.

Qualified firms obtain bid drawings and specifications for the job they are trying to win. The plans explain in detail what is expected, listing materials, dimensions, methods, and dates of completion. Each firm or joint venture tries to pull together the best possible team, compiling a list of its most reliable sub-contractors and suppliers, asking them to submit the lowest possible bids so their team wins the low bid. It's a high-stakes process as contractors introduce daring new methods, materials, and equipment to reduce their costs and lower their bids. To gain the competitive advantage, a joint venture may ask firms from all over the world to participate, ensuring that only the most proven and effective methods are used. After final calculations are made, each contractor or joint venture submits its estimate for a contract in a sealed envelope to transportation officials. If they win, they will be bound by the figures within the bid. A company's future may depend on landing one of the Big Dig's lucrative contracts.

Meanwhile, Big Dig engineers put their own "office bid" or "in-house bid" together, estimating the dollar amount of the contract that's about to go out to bid. The office bid helps the project's managers calculate the budget and schedule impacts. But it's only their best guess. Sometimes they are within a million dollars or less, sometimes they are off by tens of millions.

An official procedure, called a bid-opening, takes place in the commonwealth's transportation building. With hundreds of millions of dollars in the balance, the anticipation runs high. Under the watchful eyes of government officials, competing contractors, the public, and for a large or controversial contract, even the media, the envelopes containing the bids are opened and the dollar values are read out loud. Cell phones and beepers ignite with the good news or bad news as the bidding process comes to a climax with the announcement of the low bidder. A month or two after the winner is announced, the team receives its Notice to Proceed (NTP), the final green light needed to start the digging!

MANAGERS AND CLIENTS

The Big Dig's management consultant team is made up of two companies that were selected back in 1985, the joint venture of Bechtel and Parsons Brinckerhoff (B/PB). Overseeing the general contractors and running the administration of the daily activities of the megaproject is B/PB's primary role. They also carry out the demands of their clients, the Commonwealth of Massachusetts and the federal government, and advise them on nearly every aspect of project management. Over the last 15 years they have walked the line between consultant and business partner.

Representing us, the public, are the Commonwealth of Massachusetts and the federal government. They act as the "client," providing the cash to pay for the project, and laying out the objectives and requirements of everyone involved. In 1997 the commonwealth's legislature and executive office agreed to name the

Opposite:
Five men walk a piece of steel reinforcement bar to its final resting-place. The bar they are carrying will form the foundation of one of the highway tunnels under Boston. Over 25,000 miles of reinforcement bars will be used on the Big Dig.

Massachusetts Turnpike Authority as manager and owner of the Central Artery/ Tunnel Project. Before 1997 the Massachusetts Highway Department managed the project.

The Turnpike Authority, responsible for the stretch of I-90 from New York State to Boston, among other toll-producing entities, takes in over $500,000 daily. It was put in control of the Big Dig with the hope that it would be able to raise enough money through daily toll collections to pay for Massachusetts's portion of the projected $15,000,000,000 tab. Currently the commonwealth's contribution to the Big Dig's funding has been about 40 percent, with the federal government kicking in the other 60 percent. Before it's over, Massachusetts could be contributing as much as 50 percent of the total cost. Budget issues have been a hot debate since the Big Dig began, with intense criticism raised at the project for consistent increases in cost, and that won't change until the project is completed.

The Federal Highway Administration, the Massachusetts Turnpike Authority, and Bechtel/Parsons Brinckerhoff are independent organizations pushing to meet a collective goal. As a partnership they have managed over 40 complex design contracts and turned them into 114 megaconstruction contracts. At the peak of construction this interdependent triangle of support creates the necessary synergy and competency to see the Big Dig through to the end. It has not been easy, and many lessons have been learned.

In his book, *Rescuing Prometheus,* Thomas Hughes, a professor at the University of Pennsylvania, compared the management of the Big Dig, the Internet, the ICBM missile program, and SAGE—the first computer-aided air defense program. He dissected each project's unique structure and recognized the Big Dig for "manifesting a future trend toward an open postmodern style of coping with complexity." He observes that "those involved in the Central Artery/ Tunnel Project take into account the concerns of environmental and other special interest groups. Through public hearings, the project fosters participatory design.

CA/T [Central Artery/Tunnel] is not an elegantly reductionist endeavor; it is a messily complex embracing of contradictions." He goes on to explain that the Big Dig exemplifies a management dealing in continuous change, joint ventures, open management versus closed, and a lot of messy complexity.

MESSY COMPLEXITY

"James J. Kerasiotes is not a meek man. In a lengthy interview, he says the Governor of Massachusetts fears him. He calls the governor's chief political advisor a 'moron.' He dismisses the governor's former chief of staff, now head of the port authority, as a 'reptile.' "

The Wall Street Journal's Geeta O'Donnell Anand, led off her profile of Jim Kerasiotes on February 9, 2000, with this bold introduction. She had long been curious about the management style of the chairman of the Massachusetts Turnpike Authority, and while researching her story on the man, she became even more curious about his on-time, on-budget claims regarding the Big Dig.

The Wall Street Journal had also reported on February 2, 2000, that the Central Artery/Tunnel Project was over budget by as much as $1,400,000,000. The breaking news gripped Boston's media for months with screaming head-lines and news bulletins about billions of dollars in overruns.

President Clinton's Secretary of Transportation, Rodney Slater, was not a happy man on hearing about the budget-busting numbers. The day before the story broke, he received the state's annual financial plan for the Big Dig, documenting the project's costs. The report was still tracking the budget at $10,800,000,000 and not the reported $12,200,000,000 everyone was reading about in the papers. Secretary Slater launched his own investigation, dispatching a team of federal auditors to Boston.

Massachusetts Governor Paul Cellucci was traveling when the news broke and, like Slater, was unaware of the crisis until it hit the papers. He

The cost of the Big Dig has been a popular subject with the media since its beginning. The project has expanded in scope as well as cost since its inception in 1987, when the price was projected at $2,500,000,000.

was not happy either. State Senator Robert Havern, co-chairman of the legislature's Transportation Committee, remarked, "Was Kerasiotes hiding [the Big Dig overrun], or did he not know? Either answer is not very good. He just dropped a grenade with the pin pulled right in the governor's lap."

On Sunday, April 9, *The Boston Globe* ran the headline, U.S. AUDIT: BIG DIG IS BANKRUPT and two days later declared KERASIOTES MUST GO: "James Kerasiotes deceived the Federal Highway Administration, the Legislature, and the public about the true cost of the Central Artery project. He has performed a valuable service in getting the project more than halfway to completion, but his intellectual dishonesty has eroded people's faith in state government and ended his political usefulness to the project. . . .

"Kerasiotes's great achievement was making decisions that shaped Salvucci's vision into a politically viable project. . . . Now his most valuable contribution would be to step aside, and the questionable management practices that have led us to this story should leave with him."

The writing was on the wall. The next day *The Globe* reported that it was over: "Governor Paul Cellucci yesterday fired James J. Kerasiotes moments after the nation's top highway official accused the Big Dig chief of engaging in an 'unconscionable' betrayal of the federal government by 'intentionally concealing the project's soaring cost overruns that could reach $2 billion.'" They also reported Slater as saying that the failure to disclose cost overruns immediately "stands as one of the most flagrant breaches of the integrity of the federal-state partnership in the history of the nearly 85-year-old federal aid highway program."

"Jim did something that was really important," explained David Luberoff, a project historian at Harvard University. "He put cost on the table, and he closed out the last of the major design and permitting disputes." Kerasiotes also received praise for a remarkably good safety record and for keeping the city's traffic moving throughout construction.

Kerasiotes fought until the very end, and when it came time to step down he issued a statement saying, "I tell you, in my heart, I feel I pushed back on the budget's bottom line for only one reason: It was how I did business, to push back, to demand. Maybe I pushed too hard; maybe I demanded too much." People close to him know he wanted to see the project through to the end. In his resignation letter to Governor Cellucci, Kerasiotes wrote, "I am saddened and disappointed at this turn of events. . . . I believe when the final reckoning is made, my record will stand as a solid one in service to the people of the Commonwealth."

The Big Dig will have removed 80,000,000 wheelbarrows of earth from its construction sites by the time the project is complete.

THE BEAN COUNTER

Governor Cellucci immediately named Andrew S. Natsios as the new chairman of the Massachusetts Turnpike Authority. As chairman, Natsios automatically became responsible for the management of the Big Dig. Previously he was the commonwealth's Secretary of Administration and Finance, overseeing 22 agencies, and was responsible for the state's budget. He is one of the governor's oldest and most trusted political friends, and was no doubt picked because of his reputation as a scrupulous bean counter.

He may not be an engineer or construction manager, but he understands public finance and policy. "Natsios is so frugal with state funds that his government-issued car, a Dodge Neon, was ordered without a tape deck or air conditioning," reported *The Boston Globe*. When asked the morning after his appointment how it felt to be in charge of the largest public project in the history of the nation, his reply was, "I didn't sleep well last night." And he later added, "It's our problem and we are going to fix it."

THE FEDS KEEP DIGGING DEEPER

Natsios is going to need his sleep. On June 16, 2000, Laura Brown reported in *The Boston Herald*, "FBI agents and other federal investigators this week began grilling people linked to . . . the Big Dig in the wake of the project's skyrocketing cost overruns, sources said." And a few days later *The Wall Street Journal* reported that "The Federal Bureau of Investigation launched a criminal probe into the Big Dig cost overrun, joining a civil investigation by the Justice Department and a separate investigation by the Securities and Exchange Commission." The article went on to say that an "audit could show the financial cost of the Boston highway project will be even higher than the $13.6 billion estimated by the highway administration."

Inset: A picture of the Grain Exchange Building with the elevated Central Artery before the Big Dig. A watercolor painting of the same scene after the Big Dig. The tree-lined boulevard is Atlantic Avenue and below it is the 10-lane superhighway.

Clockwise: The Big Dig is building the largest vehicle tunnel ventilation system in the world with seven ventilation structures throughout Boston. Vent building number 7 (left) on the tarmac of Logan airport has won architectural awards for its industrial design. A technician inspects one of the 24 massive fans inside the building. A jeep drives through the Ted Williams Tunnel directly beneath vent building number 7. When the Big Dig is finished it will deliver an advanced electronic monitoring system, with over 400 video cameras, 130 electronic message signs, and 30 infrared height detectors.

It appears likely that the overrun that Kerasiotes announced at $1,400,000,000 may edge closer to $2,200,000,000. Natsios, determined to cooperate in every way, has turned over forty boxes of documents to federal investigators.

The investigation continues, and so does the Big Dig.

THE END IS NEAR

In 2005, when the work is finished, the Big Dig will deliver more than a safe and efficient highway. It will liberate Boston from the 1950s public works project that scarred the city, damaging neighborhoods, homes, businesses, and a way of life. Correcting wrongs from the past are the roots of the Big Dig; improving the region's quality of life is its promise. It's an environmental, public transit, urban planning, utility-upgrading, park-giving, public safety, harbor cleanup, historical-renovation, road-building project.

The old port town has successfully reached back to its daring maritime roots, taking risks and reaping the hard-fought rewards. Dr. William Fowler, director of the Massachusetts Historical Society, says, "I am tired of hearing people talk about what it costs! Because you know what? Thirty years from now no one will remember. But they will remember the result. To save $100,000,000 now is very poor policy. If some of the critics that we have today were around during the 1850s, you and I would be sitting here smelling mud and looking at dead rats!"

Yesterday there were traffic snarls, dirt, grime, and urban blight along the city's waterfront. Today dust, heavy equipment, and the unmistakable bustle of the Big Dig signal billions of dollars of investment in the works. Boston is already a different city from what it was when the project got under way, and tomorrow should be remarkable. Over 260 acres and $250,000,000 worth of parks and land-scaped plazas will replace the elevated highways, a city dump, and a railroad yard. Thousands of trees and broad new sidewalks will line the streets.

A 100-year struggle to build a logical local road system between North

Station and South Station finally ends with a tree-lined boulevard and acres of open space along the route. The city's dream of connecting the Charles River park system with Boston Harbor will become a reality with a link of miniparks that will turn a rusting industrial corridor into acres of trails, trees, and green space.

Boston's Fort Point Channel, the historic and still active industrial district, will become a new waterfront area, with a park, boardwalks, walkways, and nearly a mile of restored seawalls.

Across the harbor, East Boston has begun to win back its neighborhood from the invasive growth of overburdened highways and an expanding airport. The Big Dig will improve the roads and increase mass transit use with improvements to the Blue Line subway. Most importantly, the borough gets a new park and an expansion to an existing park.

In keeping with a broken tradition, Boston once again looks toward the water and sees its bright future. Just beyond the city's inner harbor, one mile from the shore of South Boston, is Spectacle Island. The once noxious, burning city dump will soon become the gateway to the harbor's new national park area. Spectacle's beaches, five miles of walking trails, a marina, and thousands of plants and trees will attract Bostonians and thousands of visitors to this reborn island.

The words of one commonwealth citizen sum it up. Says Larry Kolenberg of Boston, "This is the greatest project in the world and it's happening right here in our backyard! I've seen more than a few documentaries comparing the Big Dig to every project on the globe, past and present, and it ranks up there with the best of them. Sure there are controversies, but the Big Dig is like the Red Sox. We may bitch a little, but it's our team and we love 'em in the end."

ACKNOWLEDGMENTS

Many articles, publications, and books were helpful in my research for this book. But a special debt is owed to these works and their authors: David Luberhoff, Alan Alshuler, and Christie Baxter, *Megaprojects*; Thomas Hughes, *Rescuing Prometheus*; MIT Press, *The Mapping of Boston*; Walter Muir Whitehill, *Boston—A Topographical History*; Bill Angelo's articles in "Engineering News Record." In addition, many excellent articles in *The Boston Globe, The Boston Herald,* and *The Wall Street Journal* were helpful. A special thanks to Andrew Natsios, Chairman of the Massachusetts Turnpike Authority, for opening the doors, the windows, and the books of the Big Dig; the Joint Venture of Bechtel/Parsons, Brinckerhoff for professional assistance; Tony Lancellotti and the Resident Engineers and Field Engineers who have schooled me for years in disciplines of civil engineering; Brian Brenner, P.E., Bechtel/Parsons, Brinckerhoff for graciously contributing his expertise and time to the writing of Chapter 7, "Slurry Walls and the Maze Below"; to Senator Jack Quinlan, for his friendship, his loyal support, and his sage advice; my sister Meg and her husband, Paul Scarpetta, for their input and positive reinforcement; Laura and Andy Fisher for their guidance; Melissa Koff, for keeping the light bright and for her excellent editorial support; Dennis Rahilly and Sheila and Bob Wenstrup for technical and historical support; Lieutenant General William F. Flynn, who kept me in line and always encouraged me to write; and finally to Barbara J. Morgan, for believing in this book and for committing herself and her wonderful team of professionals, Marjorie Palmer and Richard J. Berenson to the cause.

CREDITS

APPENDICES & INDEX

THE BIG DIG TIMELINE

April 1983:
Work begins on Environmental Impact Statement/Report.

May 1985:
Bechtel/Parsons Brinckerhoff is hired as management consultant.

January 1987:
Building-acquisition and business-relocation process begins (no private homes taken).

April 1987:
Congress approves funding and scope of project. Staff and offices are established.

November 1988:
Exploratory archaeology digs begin.

November 1990:
Congress allocates $755 million to project.

January 1991:
The Massachusetts Executive Office of Environmental Affairs approves the project's Final Supplemental Environmental Impact Report (with the condition that improvements to the Charles River Crossing receive further study); geotechnical study begins.

May 1991:
Federal Highway Administration issues Record of Decision, the construction go-ahead. Advertisement of first construction contracts.

June 1991:
Commonwealth of Massachusetts signs agreement with environmental groups to pursue comprehensive parkland and transportation mitigation initiatives.

December 1991:
Construction begins on Ted Williams Tunnel.

March 1992:
Program to ensure continuous utility service during construction is announced by project and local utility companies.

July 1992:
Construction of I-90 South Boston approach to Ted Williams Tunnel begins.

August 1992:
Construction of I-90 Logan airport approach to Ted Williams Tunnel begins.

Downtown utility relocation to clear the path for upcoming Central Artery Tunnel construction begins.

September 1992:
The first of 12 Immersed Tube Tunnel sections for Ted Williams Tunnel arrives in Boston.

October 1992:
A half-mile stretch of railroad track is relocated to make way for the tunnel approach in South Boston. The work represents the first completed construction contract on the project.

November 1992:
Archaeologists find seventeenth- and eighteenth-century artifacts at a North End dig.

December 1992:
More than $1 billion in design and construction contracts underway.

February 1993:
The first ITT section of Ted Williams Tunnel is placed in Boston Harbor.

A 24-hour telephone service to answer questions on Central Artery/Tunnel construction, traffic, and commuting news is introduced.

April 1993:
Work begins in the Financial District as a 30-inch gas line is moved to make way for the underground Central Artery. Project crews continue to relocate sewer lines from Broadway to Summer Street.

May 1993:
The largest circular cofferdam in America is completed on the waterfront in South Boston. The cofferdam holds back earth and water in the area where the ITT tunnel connects to the land-based tunnel.

July 1993:
Construction begins on temporary loop ramps linking the existing Central Artery with Route 1 via new tunnels underneath City Square in Charlestown.

A Draft Supplemental Environmental Impact Statement/Report for the project's Charles River Crossing is filed.

August 1993:
The final Ted Williams Tunnel ITT arrives in Boston Harbor. Eight of the 12 steel-and-concrete tubes are placed and connected in the trench on the harbor floor.

History of Budget Increases	
1983	$2.6 billion
1987	3.2 billion
1990	4.4 billion
1991	5.8 billion
1993	6.5 billion
1994	7.8 billion
1996	10.8 billion
2000	13.6 billion

November 1993:
The Central Artery/Tunnel Project reaches its most significant milestone to date with the placement of the final two ITT sections for Ted Williams Tunnel.

March 1994:
Project receives certificate from the Massachusetts Executive Office of Environmental Affairs, giving final state environmental approval for the new Charles River Crossing design.

June 1994:
Project receives Record of Decision from the Federal Highway Administration, completing the federal environmental review process for the new Charles River Crossing.

September 1994:
Temporary loop ramps linking I-93 with Route 1 open, eliminating the dangerous traffic weave over the Charles River. The opening of the ramps completes the Massachusetts Highway Department's Central Artery North Area (CANA) project.

November 1994:
The Rapid Service Press Building is razed in preparation for I-93 Interchange.

February 1995:
Massachusetts Highway Department awards two construction contracts totaling $524 million to construct major sections of the new underground Central Artery. The awards bring to $2.1 billion the value of the awarded contracts on the project.

August 1995:
Construction begins for the underground expressway beneath Atlantic Avenue and Kneeland and Congress Streets.

December 1995:
Ted Williams Tunnel opens to commercial traffic.

October 1999:
Charles River Crossing Storrow Drive bridge opens.

2002:
Spectacle Island park to open.

I-90 extension to Ted Williams Tunnel to open. Tunnel will be open to all traffic.

Northbound lanes of Central Artery and 10-lane cable-stayed bridge over Charles River will open to traffic.

2003:
Southbound lanes of Central Artery and cable-stayed bridge will open to traffic.

2004:
Demolition of elevated highway to finish.

2005:
Project complete.

CENTRAL ARTERY/TUNNEL PROJECT PARTICIPATING UNIONS

Asbestos Workers –
Local Union No. 6
Dorchester, MA

Boilermakers –
Local Lodge No. 29
North Quincy, MA

Boston Building Trades
Council
Dorchester, MA

Bricklayers –
Local Union No. 3
Charlestown, MA

Building and Construction
Trades Department, AFL-CIO
Washington, DC

New England Regional Council
of Carpenters
South Boston, MA

Carpenters –
Local Union No. 33
Boston, MA

Carpenters –
Local Union No. 40
Cambridge, MA

Carpenters –
Local Union No. 51
South Boston, MA

Carpenters –
Local Union No. 56
South Boston, MA

Carpenters –
Local Union No. 67
Dorchester, MA

Carpenters –
Local Union No. 218
Medford, MA

Carpenters –
Local Union No. 1121
Allston, MA

Carpenters –
Local Union No. 2168
South Boston, MA

Electrical Workers –
Local Union No. 103
Dorchester, MA

Elevator Constructors –
Local Union No. 4
Allston, MA

Iron Workers –
Local Union No. 7
South Boston, MA

Laborers' District
Council of Massachusetts
Hopkinton, MA

Laborers' Local
Union No. 22
Malden, MA

Laborers' Local
Union No. 88
Quincy, MA

Laborers' Local
Union No. 151
Cambridge, MA

Laborers' Local
Union No. 223
Dorchester, MA

Laborers' Local
Union No. 1421
Boston, MA

Massachusetts State
Building and Construction
Trades Council
Dorchester, MA

Operating Engineers –
Local No. 4
Roslindale, MA

Operating Engineers –
Marine Division Off-Shore
Dredging & Drilling Operations
– Local No. 25
Metuchen, NJ

Painters District
Council – No. 35
Roslindale, MA

Pipefitters –
Local Union No. 537
Allston, MA

Plasterers & Cement Masons –
Local Union No. 534
Boston, MA

Plumbers –
Local Union No. 12
Boston, MA

Roofers –
Local Union No. 33
Dorchester, MA

Sheet Metal Workers –
Local Union No. 17
Dorchester, MA

Sprinkler Fitters –
Local Union No. 550
Boston, MA

Teamsters –
Local Union No. 379
South Boston, MA

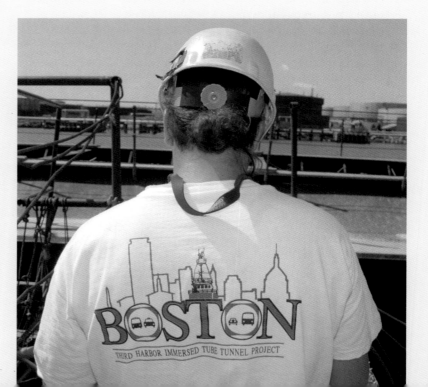

INDEX

Page numbers in *italics* refer to illustrations and captions.

A

Abbot, Kathy, 101
Abrahms, Sally, 35
Alger, Cyrus, 178
Anand, Geeta O'Donnell, 223
Antonelli, Leo, 136
archaeologists, 234
    at Boston University, *151*
    in Paddy's Alley, 147
    on Spectacle Island, 86, *150*
Associated Press, 131
Atlantic Avenue, *128–29, 130–31*,
    131
    utility lines under, *16*
augers, *78–79*

B

Baby Bridge. *See* Storrow Drive
    Connector Bridge
Bales, John, 186, 188, 191
Banker & Tradesman, 210
barges, 190
Barnhart, Ray, 38
Bechtel Corporation, 42, 221
Bechtel/Parsons Brinckerhoff, 221,
    222, 234
    and Fort Point Channel
        crossing, 174
    previous projects, 42
Bergnazzani, Pete, *160–61*
Berkowitz, Glen, 210
Bertoulin, Mike, 159, 173, 191, 195
Bethlehem Steel, 56, *64*, 179
bidding process, 218, 221
"Big Dig Sinking Downtown
    Skyscraper" *(Boston Herald)*, 131
Bill, Samuel, 87
Black Falcon Terminal, 63, 70
blasting in the harbor, 61–63,
    62–63
Blue Line subway, *11*, 22, 54, 120

Boston
    early settlement of, 87–88
    first subway system, 20–21, 54
    man-made land in, 18–19,
        20–21
    map, *12*
    wharves, 19–20
Boston Chamber of Commerce, 144
Boston Fire Department, 153
Boston Globe, The,
    34, 36, 147, 160, 225, 226
    on cost of Big Dig, 224, 225
    on Fort Point Channel crossing,
        174
    on Harbor Islands, 101
    on rats, 49
Boston Harbor, 231
    blasting in, 61–63, *62–63*
    contaminated material from,
        44, 57–58, 60
Boston Herald, The,
    31, 39, 131, 132, 226
Boston Redevelopment Authority,
    32
Boston Transit Authority, 21
Boston University, *151*
Boston Wharf Company, 178
bottom-up construction, 138
Boylan, Mike, 112
B/PB. *See* Bechtel/Parsons
    Brinckerhoff
Bridge Design Review Committee,
    201
Brown, Laura, *31*, 226
Brown and Rowe, 97, 100
    landscaping plan for
        Spectacle Island, *102–3*
budget
    to alleviate disruption,
        14, *142–43*
    cost of project, *224*
    increases, 223, 235

building settlements,
    instruments measuring, *132*
Burden, George, 18
business relocations, 234
Button, John, 18
Byrd, Robert, 38

C

cable-stayed bridges, 202–3
Callahan, William,
    22, *22*, 25, 27, 32, 54
Callahan Tunnel, 30, 54
Cambridge, 45, 202
Campbell, Robert, 201
Carnes, John, *149*, 152
Cashman. *See* Perini, Kiewit and
    Cashman
CBS-TV, *46*
Cellucci, Paul, 223, 225, 226
Central Artery,
    13, *15*, 22–23, *26*, 27, *227*
    demolition of, 235
    destruction caused by,
        23, 23–24
    traffic on, *24–25*, 28
    underpinning, 140
Chandra, Vijay, 202
Charles River, 10, 200
    park system, 231
Charles River Bridge, 125, 200
Charles River crossings,
    45, 198–213
    bridge, 204, *207, 211*
    computer image, *212–13*
Charlestown, 18, 45, 204
Chinatown, elevated highway
    through, 23–24
Christian Menn and Company, 213
Christian Science Monitor, The, 215
Chunnel, 42, 168
cofferdams, 80, 82
Colvin, Bruce, 48, 49

contaminated material
    brought to Governors Island, 58
    handling of, *44*
    from harbor floor, 57–58, 60
Crane Farm. *See* South Bay
    Interchange
cranes, *122–23, 128–29*, 135
Curley, James Michael, 22

D

deep-soil-mixing, 175, *176*
Deer Island, 101
dewatering campaign, 131
Dewey Square Tunnel. *See* South
    Station Tunnel
divers, 190–91
    and construction of
        Ted Williams Tunnel, 73
Dole, Elizabeth, 37–38
Dole, Robert, 39
Dorchester, 18
downtown construction,
    118–53, *140–41*
dredging operations,
    56, *57–59*, 186
Dukakis, Michael,
    29, 29–30, 34, 35, 36, 45, 83
    and Spectacle Island, 90
Durlacher, Stan, 201

E

earth moving, 93, *140–41*, 225
    to Spectacle Island,
        86, 93, *94–95, 96–99*
East Boston, 231
Eisenhower, Dwight D., 10
electric lines, 107
electronic monitoring system,
    228–29
Ells, Steve, 201
employment, 13, 217, *218–19*

Engineering News-Record, 42, 158
environment, 43, 234, 235
    fish protection, 61–63, *62–63*
    sustainability of
        Spectacle Island, 102
Environmental Impact Report (EIR),
    43, 234
Environmental Impact
    Statement (EIS), *41*, 45, 234

F

Federal Aviation Administration
    (FAA)
    and construction of
        Ted Williams Tunnel, 63
    and permits, 49
federal funding, 28, 29, 38–39
Federal Highway Administration,
    36, 222
    and crossing of the
        Charles River, 201
fire protection requirements, 153
fish-monitoring program, 61
flat jacks, 140
Flynn, Raymond, 46, 48
Flynn, William F., 50, 155, 156, 158
Fort Point Channel, 231
    bridges over, 177–78
    casting basin, *182–89, 191–93*
    contaminated material
        from, *44*
    crossing, 172–97
    dredging operations, 186
    stabilizing weak soil, *172–73*
Foster, John, 157–58
Fowler, William, 230
Frank, Barney, 36
Franzen, Al, 180
Freedom Trail, 145
Friends of Boston Harbor
    Islands, 92

G

Gillette, 179, 184
Global Positioning Systems (GPS),
    70, 194
Goldberg, Carey, 90

Golden Gate bridge, 203
Goodrich, Bill, 139
Governors Island, 58
"Green Monster, The."
    See Central Artery
ground movement vibrations,
    instruments measuring, 132

H

Hancock, John, 200
Harrington, Kevin, 30
Harris, "Big Jim," 190, 191, 194
Hasenstab, Bob, 147
Havern, Robert, 224
Herter, Christian, 24
High Bridge, 210
Hill, John, 18–19
Hughes, Thomas, 222

I

Immersed Tube Tunnel (ITT)
    sections
    GPS used in placing, 70, 194
    used for Fort Point Channel
        crossing, 174, 179–80, *180*,
        183–84
    used for Ted Williams Tunnel,
        55, 56, *56–57*, 60, 63–75, *64,*
        66, 67, 69–71, 74–75,
        234, 235
    working on, *70–73, 76–77*
Inner Belt, 29
Interbeton. *See* Slattery, J.F. White,
    Interbeton and Perini
Interim Viaduct over Albany Street
    (IVAS), 158
International Union of Operating
    Engineers, 159, *161*
Interstate-90,
    10, 32, *156–57*, 195, *195–97*
Interstate-93, 10, 22, 120, 135,
    146, *154, 156–57, 204–5*
Interstate-95, 29
Interstate Highway System, 13
IVAS. See Interim Viaduct over
    Albany Street

J

J.F. White. *See* Slattery, J.F. White,
    Interbeton and Perini
John F. Fitzgerald Expressway,
    147, 152, 153
Joint Venture. *See* Bechtel/Parsons
    Brinckerhoff
joint ventures, 218

K

Kennedy, Edward, 38
Kerany, Rachid, *123*
Kerasiotes, James,
    202, *202*, 210, 223, 225, 230
"Keriasiotes Must Go"
    (Boston Globe), 224
Kiewit. *See* Perini, Kiewit and
    Cashman
King, Edward J., 30, 34, 83
Kolenberg, Larry, 231

L

landmaking in Boston,
    18–19, *20–21*
landscaping of Spectacle Island,
    97, 100, *102–3*
Larsen, John, 158
Larson, Tom, 201
Layton, Lyndsey, 135
Lewis, Mike, 174, 175
Logan International Airport, 10
    and Silver Line, 128
    and Ted Williams Tunnel, 54, 77
Luberoff, David, 225

M

MacDonald, Brian, 86
Marinellia, Jim, 217
Massachusetts Bay Disposal Site,
    58, 60
Massachusetts Department of
    Public Works, 41
Massachusetts Executive Office
    of Environmental Affairs,
    43, 234, 235

Massachusetts Highway
    Department, 222, 235
Massachusetts Historical Society,
    230
Massachusetts Port Authority
    (Massport), 58
    and Big Dig, 77, 78
Massachusetts Turnpike
    Authority, 222
McGee, Marianne, 146
McGrath, Lauren, *145*
Menn, Christian, 203, 213
Metrowest Daily News, The, 33
Mill Creek, 19
milling machine, *116–17*
Mill Pond, 19, 147
Modern Continental and
    Weidlinger Associates, 140, 158
Mohler, Paul, 151

N

National Environmental Policy Act,
    50
National Historic Preservation Act,
    86
Natsios, Andrew S., 226, 230
Naylor, Edward, 152
Naylor, Katherine Nanny, *149*, 152
Nee, Frank, 138
New York Times, The, 88, 90
Nordell, John, 215
North End, elevated highway
    through, 23–24
North Station, 230
Noyes, Oliver, 19

O-P

Obenshain, Greg, 165
O'Neill, Thomas "Tip," Jr.,
    24, *36,* 37, 38
Orange Line subway, 22
Outstanding Civil Engineering
    Achievement Award, 53
Paddy's Alley, 147, *148–49*
Parsons, Brinckerhoff, Quade &
    Douglas, 42–43, 221
Patriot Ledger, The, 135, 136

Paul, Jeffrey M., 49, 50
Perini. *See* Slattery, J.F. White,
    Interbeton and Perini
Perini, Kiewit and Cashman (PKC)
    joint venture, 126
permits, 49–50, *50–51*
pest control companies, 48–49
Poison Pit of Death, 60
Powell, Nancy, 131

Q-R

Quinlan, Jack, 60
Ramp-D, jacking of,
    164–71, *166–67, 170–71*
rat scare, 46–49
Reagan, Ronald, 37, 38, 39
Red Line subway, 22, 120, 135
    bridge for, *134*, 136
Rescuing Prometheus (Hughes), 222
Reserve Channel, 63
Reynolds, Bill, 33, 34, 35
Robertson, Jack, 144, 147
Rodent Control Department, 46, 49
Rodwell, Jason, 162, 168
Rosenthal, David, 116
Rumney Marsh, 181
Ryan, Bill, 159
Ryan, Mike, 153
Ryder, Mike, 161

S

safety record, *132–33, 218–19*
Sailors, David, 179
Saltonstall, Leverett, 22
Salvucci, Frederick P.,
    30–36, *31*, 41–43, 46, 48, 125
    and crossing of the
        Charles River, 200
    and Ted Williams Tunnel, 55
sandhogs, 136, *137, 163, 168–69*
Sanford, Terry, 38–39
Sargent, Frank, 29, 34
Save the Harbor, 92
Scheme Z, 45, 200–203
Sève, Peter de, *46–47*
sewer lines, 108

Silver Line Transitway,
    126, 128, *134*, 135, *195–97*
Slater, Rodney, 223, 225
Slattery, J. F. White, Interbeton
    and Perini joint venture,
    146, 164, 168
slurry walls, 105–17
    construction, *104–5*, 109–16,
        110–13, 116–17
    defined, 106
Smith, Bill, 160–61
Smith, John, 200
soil stabilization, *172–73*, 175–77
Soletanche Company, *114*
South Bay Interchange,
    154–71, *156–57*, 178
South Station, 21, 178
    link to North Station, 231
    tunnel under, 24, 162
Spectacle Island, 85–103, *89*, 231
    archaeologists on, *150*
    contaminated material from, *44*
    earth brought to,
        86, 93, *94–95, 96–99*
    garbage on, *91*
State Street, 120
steam pipes, 107
Storrow Drive Connector Bridge,
    199, *204–5*, 206, *208–10*
Subaru Pier, 60, *96*
Summer Street, 131
Sumner Tunnel, 30, 54, 125
Super Scoop, 56, 60, 63
Surface Transportation and
    Uniform Relocation Assistance
        Act of 1987, 38
Symons, Henry, 19

T

tagmen, 159
Tampa Steel Erecting Company, 206
Ted Williams Tunnel, *13*, 53–83
    changes in design of, 55–56
    construction, *13*
    construction of East Boston
        approach, 77–79
    construction of marine tunnel,
        56–75

construction of South Boston
    approach, *80–81*, 80–82
contracts for, 56
fill from used on
    Spectacle Island, 86
ventilation building, 82
temporary roadways, *142–43*
timeline, 234–35
Time magazine, *46*
Tobin, Maurice, 22
top-down construction, 138
tower cranes, 161
tunnel-jacking operation,
    161–62, *163*

U-V

underground expressway.
    See downtown construction
unions, participating, 236
United States Postal Service, 178
U.S. Army Corps of Engineers, 58
    and permits, 49
    and Spectacle Island, 92
"U.S. Audit: Big Dig Is Bankrupt"
    (Boston Globe), 224
U.S. Coast Guard
    and construction of
        Ted Williams Tunnel, 63, 70
    and permits, 49
U.S. Interstate Act, 13
utility lines, 106–9
    phone lines drowned, 144
    relocation of, *107–9*
vehicle ventilation system, *228–29*
Virta, Mike, 95, 101
Volpe, John, 24, 25

W-Y-Z

Wall Street Journal, The,
    46, 223, 226
and rat scare, 46
Ward, Nahum, 88
Washington Post, The, 38
Weld, William, 54, 56, 83, *83*
White, Kevin, 31, 32–33, 34
Whittinghill, Jim, 39
Williams, John Henry, *83*

Williams, Ted, 54, 83, *83*
Winthrop, John, 18
Wirth, Jacob, 33
Wood, Sam, 46
workers, *124–25*.
    See also sandhogs; tagmen
    safety, *132–33, 218–19*
Wright, James, 38
Yankee magazine, *46–47*
Zuk, Peter M., 10, *41*, 49